SpringerBriefs in Applied Sciences and Technology

SpringerBriefs present concise summaries of cutting-edge research and practical applications across a wide spectrum of fields. Featuring compact volumes of 50–125 pages, the series covers a range of content from professional to academic.

Typical publications can be:

- A timely report of state-of-the art methods
- An introduction to or a manual for the application of mathematical or computer techniques
- A bridge between new research results, as published in journal articles
- A snapshot of a hot or emerging topic
- An in-depth case study
- A presentation of core concepts that students must understand in order to make independent contributions

SpringerBriefs are characterized by fast, global electronic dissemination, standard publishing contracts, standardized manuscript preparation and formatting guidelines, and expedited production schedules.

On the one hand, **SpringerBriefs in Applied Sciences and Technology** are devoted to the publication of fundamentals and applications within the different classical engineering disciplines as well as in interdisciplinary fields that recently emerged between these areas. On the other hand, as the boundary separating fundamental research and applied technology is more and more dissolving, this series is particularly open to trans-disciplinary topics between fundamental science and engineering.

Indexed by EI-Compendex, SCOPUS and Springerlink.

More information about this series at http://www.springer.com/series/8884

João P. S. Rosa · Daniel J. D. Guerra ·
Nuno C. G. Horta · Ricardo M. F. Martins ·
Nuno C. C. Lourenço

Using Artificial Neural Networks for Analog Integrated Circuit Design Automation

 Springer

João P. S. Rosa
Instituto de Telecomunicações, Instituto
Superior Técnico
University of Lisbon
Lisbon, Portugal

Daniel J. D. Guerra
Instituto de Telecomunicações, Instituto
Superior Técnico
University of Lisbon
Lisbon, Portugal

Nuno C. G. Horta ⓘ
Instituto de Telecomunicações, Instituto
Superior Técnico
University of Lisbon
Lisbon, Portugal

Ricardo M. F. Martins
Instituto de Telecomunicações, Instituto
Superior Técnico
University of Lisbon
Lisbon, Portugal

Nuno C. C. Lourenço
Instituto de Telecomunicações, Instituto
Superior Técnico
University of Lisbon
Lisbon, Portugal

ISSN 2191-530X   ISSN 2191-5318   (electronic)
SpringerBriefs in Applied Sciences and Technology
ISBN 978-3-030-35742-9   ISBN 978-3-030-35743-6   (eBook)
https://doi.org/10.1007/978-3-030-35743-6

This Springer imprint is published by the registered company Springer Nature Switzerland AG
The registered company address is: Gewerbestrasse 11, 6330 Cham, Switzerland

To João, Fernanda and Filipa
João P. S. Rosa

To Carla, João and Tiago
Nuno C. G. Horta

To Martim
Ricardo M. F. Martins

To Alina, Íris and Ana
Nuno C. C. Lourenço

Preface

In the last years, the world has observed the increasing complexity of integrated circuits (ICs), strongly triggered by the proliferation of consumer electronic devices. While these ICs are mostly implemented using digital circuitry, analog and radio frequency circuits are still necessary and irreplaceable in the implementation of most interfaces and transceivers. However, unlike the digital design where an automated flow is established for most design stages, the absence of effective and established computer-aided design (CAD) tools for electronic design automation (EDA) of analog and radio frequency IC blocks poses the largest contribution to their bulky development cycles, leading to long, iterative, and error-prone designer intervention over their entire design flow. Given the economic pressure for high-quality yet cheap electronics and challenging time-to-market constraints, there is an urgent need for CAD tools that increase the analog designers' productivity and improve the quality of resulting ICs.

The work presented in this book addresses the automatic sizing and layout of analog ICs using deep learning and artificial neural networks (ANNs). Firstly, this work explores an innovative approach to automatic circuit sizing where ANNs learn patterns from previously optimized design solutions. In opposition to classical optimization-based sizing strategies, where computational intelligence techniques are used to iterate over the map from devices' sizes to circuits' performances provided by design equations or circuit simulations, ANNs are shown to be capable of solving analog IC sizing as a direct map from specifications to the devices' sizes. Two separate ANN architectures are proposed: a regression-only model and a classification and regression model. The goal of the regression-only model is to learn design patterns from the studied circuits, using circuit's performances as input features and devices' sizes as target outputs. This model can size a circuit given its specifications for a single topology. The classification and regression model has the same capabilities of the previous model, but it can also select the most appropriate circuit topology and its respective sizing given the target specification. The proposed methodology was implemented and tested on two analog circuit topologies.

Afterward, ANNs are applied to the placement part of the layout generation process, where the position of devices is defined according to a set of topological constraints so that the minimum die area is occupied and the circuit's performance degradation from pre-layout to post-layout is minimized. Analog IC layout placement is one of the most critical parts of the whole circuit design flow and one of the most subjective as well, as each designer has his/her own preferences and layout styles when placing devices. A model using ANNs is trained on previous placement designs, and instead of explicitly considering all the topological constraints when doing this process, the ANN learns those constraints implicitly from the patterns present in those legacy designs. The proposed model takes as input the sizing of the devices and outputs their coordinates in the circuit layout. The ANNs are trained on a dataset of an analog amplifier containing thousands of placement solutions for 12 different and conflicting layout styles/guidelines and used to output different placement alternatives, for sizing solutions outside the training set at push-button speed.

The trained ANNs were able to size circuits that extend the performance boundaries outside the train/validation set, showing that, more than a mapping for the training data, the model is actually capable of learning reusable analog IC sizing patterns. The same is valid for the placement model, which not only replicates the legacy designs' placement, but also shows indications that it learns patterns from different templates and applies them to new circuit sizings. Ultimately, both methodologies offer the opportunity to reuse all the existent legacy sizing and layout information, generated by either circuit designers or EDA tools.

Finally, the authors would like to express their gratitude for the financial support that made this work possible. The work developed in this book was supported by FCT/MEC through national funds and when applicable co-funded by FEDER–PT2020 partnership agreement under the project UID/EEA/50008/2019.

This book is organized into five chapters.

Chapter 1 presents an introduction to the analog IC design area and discusses how the advances in machine learning can pave the way for new EDA tools.

Chapter 2 presents a study of the available tools for analog design automation. First an overview of existing works where machine learning techniques are applied to analog IC sizing is presented. Then, the three major methodologies used in the automatic layout of analog ICs are described.

Chapter 3 conducts a review of modern machine learning techniques, where the advantages and disadvantages of several machine learning methods taking into account its application to the automation of analog integrated circuit sizing and placement are discussed. Afterward, a brief overview of how the ANN learning mechanism works, the optimization techniques to speed up the convergence of the learning algorithm and the regularization techniques are presented.

Chapter 4 presents two ANN models for analog IC sizing, i.e., a regression-only model and a classification and regression model. The first serves as a proof of concept of the applicability of ANNs to analog sizing and the second that selects the most appropriate circuit topology and respective sizing for a target specification.

Chapter 5 introduces the exploratory research using ANNs to automate the placement task of analog IC layout design. The ANNs are trained on a dataset of an analog amplifier containing thousands of placement solutions for 12 different and conflicting layout styles/guidelines and used to output different placement alternatives, for sizing solutions outside the training set at push-button speed.

Lisbon, Portugal
João P. S. Rosa
Daniel J. D. Guerra
Ricardo M. F. Martins
Nuno C. C. Lourenço
Nuno C. G. Horta

Contents

Abbreviations

Adam	Adaptive Moment Estimation
AI	Artificial Intelligence
AMS	Analog and/or Mixed-Signal
ANN	Artificial Neural Network
CAD	Computer-Aided Design
CMOS	Complementary Metal-Oxide-Semiconductor
CNN	Convolutional Neural Network
EA	Error Accuracy
EDA	Electronic Design Automation
ELU	Exponential Linear Unit
EOA	Error and Overlap Accuracy
GBW	Gain–Bandwidth
IC	Integrated Circuit
IDD	Current Consumption
MAE	Mean Absolute Error
MAED	Mean Absolute Error per Device
ML	Machine Learning
MLP	Multi-Layer Perceptron
MSE	Mean Squared Error
NAG	Nesterov Accelerated Gradient
NLP	Natural Language Processing
OA	Overlap Accuracy
PM	Phase Margin
RMS	Root-Mean-Square
RNN	Recurrent Neural Network
SGD	Stochastic Gradient Descent
SoC	System-on-a-Chip
SSCE	Sparse Softmax Cross-Entropy
SVM	Support Vector Machine

List of Figures

List of Tables

Chapter 1
Introduction

1.1 Analog Integrated Circuit Design Automation

In recent years, the electronics industry has seen a tremendous increase in demand for complex and highly integrated systems that are built on a single chip for power and packaging efficiency. In the era of portable devices, integration and power consumption matter more than ever and developers are faced with the challenge of increasing the capabilities of the systems while ensuring they can be effectively integrated into energy-efficient, small, and light end products. The complexity of these systems is highly associated with the trade-off between their analog and digital sections. Developing these analog and mixed-signal (AMS) systems-on-chip (SoC) constitutes a great challenge both to the designers of chips and to the developers of the computer-aided design (CAD) systems that are used during the design process.

While in most AMS SoCs the area occupied by the digital blocks is larger than the area occupied by the analog blocks, the effort to implement the latter is considerably larger, as illustrated by Fig. 1.1.

This imbalance in the design effort, as well as the economic pressure, has motivated the development of new methods and tools for automating the analog design process, not only to reduce the costs of making those circuits (by reducing manpower), but also to reduce the time spent on design cycles. However, and despite the considerable evolution verified in the past few years, there are not a lot of EDA tools established in the industry for analog circuits compared to their digital counterpart. Analog tools to handle design centering and process variability are a notable exception, but in general, as EDA tools targeting analog ICs have not yet reached a desirable state of maturity. The design of analog ICs is time-consuming and error-prone, becoming a bottleneck to the whole design flow of a SoC [1].

This discrepancy between analog and digital automation derives from the more knowledge-intensive, more heuristic, and less systematic type of approach required for effective analog IC design. Human intervention during all phases of the design process is a constant. It is necessary to explore the solution space (both sizing of

J. P. S. Rosa et al., *Using Artificial Neural Networks for Analog Integrated Circuit Design Automation*, SpringerBriefs in Applied Sciences and Technology, https://doi.org/10.1007/978-3-030-35743-6_1

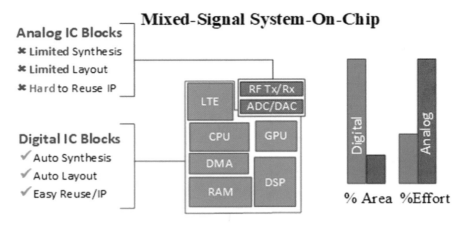

Fig. 1.1 Contrast between analog and digital blocks' design effort [1]

the devices and placement/routing to achieve the desired performances), which is enormous due to the variety of topologies and diversity of devices' types, sizes, and shapes [2]. Even with circuit simulators, layout editing environments, and verification tools, the time consumed by this task does not diminish due to the technological challenges imposed by deep nanometric integration technologies. Moreover, as the number of devices in a circuit increases so does the complexity of the design process. This is aggravated not because of the number of devices per se, but because of the number of interactions between them. More, due to the nature of the signals being handled parasitic disturbances, crosstalk and thermal noise are determinant factors for the performance of analog circuits.

All things considered, designing an analog block takes a considerably longer time than designing a digital one, and it is crucial that the designer is experienced and knowledgeable [2]. These adversities can be summarized in the following three sentences:

- Lack of systematic design flows supported by effective EDA tools;
- Integration of analog circuits using technologies optimized for digital circuits;
- Difficulty in reusing analog blocks, since they are more sensitive to surrounding circuitry and environmental and process variations than their digital counterpart [1].

1.2   Analog IC Design Flow

While the exact flow and toolchain can vary considerably between designers or companies, the majority follows the guidelines of the design flow described by Gielen and Rutenbar in [3], which generalizes the steps that designers take when manually

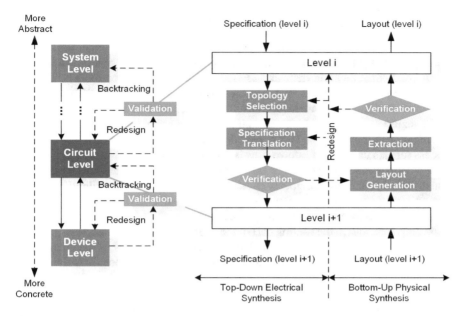

Fig. 1.2 Hierarchical levels and design tasks of analog design flow [1]

designing an analog. This design flow for AMS ICs is presented in Fig. 1.2 and is followed by most of the works on analog design automation recently developed. It consists of a series of top-down design steps repeated from the system level to the device level, and bottom-up layout generation and verification.

A hierarchical top-down methodology is useful to perform system architectural exploration since it is easier to get a better overall system optimization at a higher level and then start the more detailed implementation at the device level. This way, it is possible to identify problems early in the design process and increase the chance of first-time success, decreasing the time needed to conclude the whole process [4]. This is not always possible as the impact of device parasitics and process variations forces the execution of many iterations in real-world designs.

The number of hierarchy levels depends on the complexity of the system being handled, but the steps between any two hierarchical levels are:

- Top-down electrical synthesis path that includes topology selection, specification translation (or circuit sizing at the lowest level), and design verification;
- Bottom-up physical synthesis path that includes layout generation and detailed design verification (after layout extraction).

Topology selection corresponds to the process of selecting the most appropriate circuit topology that meets a set of given specifications at the current hierarchy level. The topology can be either chosen from a set of available topologies or synthesized.

After selecting a topology, it is needed to map the high-level block specifications into individual specifications for each of the sub-blocks. This task is called specification translation. The blocks on the lowest level correspond to single devices and this task is reduced to circuit sizing, one of the aspects of the design flow addressed in this work. Specification translation is verified using simulation tools before proceeding to the next hierarchy level. There is no device-level sizing available at higher levels, so simulations at these levels must be behavioral. In the lowest level, however, device sizing is available which makes it possible to use electrical simulations. The specifications for each of the blocks pass down to the next hierarchical level, and the process repeats until the top-down flow is completed.

Layout generation is the task of the AMS design flow that both lays the devices, whose dimensions were previously determined for the selected topology, out in the chip and connects them. It is common to split this whole process into two smaller problems: floorplanning and routing. In the floorplanning task, the devices are placed in the chip area, and in the routing part of the problem, the interconnections between devices are established. This first problem, also called placement, is the other aspect of the design flow focused in this work. It is a complex task since there are several requirements that must be considered in order to reduce the unwanted effects of parasitic currents, process variations, and different on-die operating conditions. In parallel to trying to satisfy these placement constraints, several objectives should also be minimized, such as chip area and interconnect length.

1.3   Machine Learning and Analog IC Sizing and Placement

Keeping up with the demands brought by the technological revolution is of paramount importance. EDA tools and methodologies to improve analog IC design are still an open research topic, and the developments made in ML will definitely open new research directions for EDA tools.

ML is the process in which a computer improves its capabilities through the analysis of past experiences. This area of artificial intelligence (AI) has been chosen to solve problems in computer vision, speech recognition, natural language process, among others because it can be easier to train a system showing examples of what the output should be, given a certain input than to anticipate all possible responses for all inputs [5].

Over the last two decades, ML has been greatly developed, jumping from theoretical curiosity to methods that are commercially usable. ML is slowly taking over our lives. We just don't know it yet at a conscious level. Most of the devices and services we use employ some form of a learning algorithm. Companies relentlessly mine data from their users to sell more and better products. Politicians use data from voters to manipulate their perceptions. Demand for data is at an all-time high. This development was caused, partly, by the increase in the quantity of data available and the increase in computational power available, since the performance of software using ML is directly correlated to the amount of data used to "teach" the program.

A good example of an everyday task where one can come across ML would be searching for a book or a song name in Google. Based on that search, Google will attempt to build a profile and show specific ads that target your interests. Similarly, Facebook collects massive amounts of data from its users every day, building models from the posts they make, the news they share, and the pages they like. While Facebook tells us that these models are used to improve our experience using this social network, the truth is some of the data can be used unduly and even influence elections [6]. Netflix meticulously evaluates the movies and TV shows you watch, so it can build a list of recommendations based on the shows you rated the highest. Uber not only uses ML to improve their customer experience (by training their drivers to offer customers a more comfortable experience) but also to build their own autonomous car using a model based on neural networks. Less malicious uses can be felt across computer science and across a range of industries concerned with data-intensive issues, such as drug testing in medicine, online fraud detection, and the control of logistics chains [5].

While the use of models to predict outputs from new inputs is almost instantaneous and is quite cheap, the process of automatically improving behaviors through experience can be costly and time-consuming, especially when large amounts of data must be interpreted and processed. There is a wide variety of techniques that have evolved through the years, and choosing the correct one can sometimes be a challenging task itself. One of the most used nowadays are artificial neural networks (ANNs), and those are the models considered in this work. This technique has been cyclically picked-up and abandoned over the years, but a new trend emerged recently, called *deep learning*, where much more complex networks are employed, yielding very interesting results in image processing [7], for example. Bayesian networks are used extensively for speech recognition [8]. Support vector machines can be used to classify images or to recognize handwriting [9]. Decision trees have been used for drug analysis [10]. And evolutionary algorithms have been used in several applications, namely in analog IC design [1, 2].

The fact that ML methods have been developed to analyze high-throughput experimental data in novel ways makes them an attractive approach when tackling more demanding and complex problems. Nevertheless, as powerful as these techniques may seem, at first sight, there are some caveats when trying to use them. The problem lies in the amount of the data we can collect from the situations we are trying to analyze and extract knowledge from. ML algorithms are very dependent on the number of available examples. If the problem we have at hand is hard to characterize and offers a low amount of explorable data, it will be difficult to build a model from which we can extract a set of rules that will be used to obtain the best possible generalization. Generalization is one of the key goals of every ML technique. It means that models should be trained in order to correctly classify unseen data while avoiding being too specific for the examples that were used to train the algorithm (a problem commonly known as *overfitting*).

Still, recent achievements in AI and ML and their rising success in different fields of study indicate that the knowledge mining route might bring new results for the analog IC design automation.

Analog IC performance evaluation is well established in the design flow. This prominence of circuit analysis tools and methods leads to simulation-based optimization as the most common method in both industrial [11, 12] and academic [13, 14] environments. These methods for automatic analog/RF IC sizing aim at finding the best set of device sizes by iterating over tentative guesses and their impact on circuit performance. This process is shown to be able to produce usable designs, but it is still slow, and reuse usually involves new optimization runs. Some approaches have been made to use ML algorithms in analog IC sizing. Particularly, surrogate models have been employed in this field of study, mostly as a replacement for the circuit simulator [15, 16]. Regarding automatic placement task of AMS IC layout synthesis, the research community proposals follow three main methodologies, distinguished by the amount of legacy knowledge that is used during the generation: layout generation considering placement and routing constraints (without considering legacy designs), layout migration and retargeting from a previous legacy design or template, and layout synthesis with knowledge mining from multiple previously designed layouts. However, none of them succeeded in the industry.

Therefore, applying knowledge coming from AI seems like a very auspicious way to improve analog IC design automation. This work considers the automation of two design tasks using ANNs:

- Circuit Sizing: The task of finding the devices' sizes, such as widths and lengths of transistors, resistors, and capacitors, for a given topology and technology from a set of specifications; we will also explore how ANNs can be used to select a circuit topology and correspondent devices' sizes from the set of specifications.
- Placement: The task of finding the devices' locations on the chip floorplan given their dimensions. In this work, ANNs are used to create a model that, given the sizes of the devices of a circuit, outputs their layout position in the IC. The proposed ANN also outputs three different placements (one with the smallest layout area, one with the minimum aspect ratio, and one with the maximum aspect ratio) to be used depending on the situation.

These models can be used to reduce the need of human intervention, automating part of the process, and in conjunction with each other, designers would have only to adjust minor details in the final circuit proposed by the models. Another appeal of using an ML system is the possibility of reusing circuits that were designed before. There are already a huge amount of analog layout designs stored that can be used as a baseline for new layouts. ML should accomplish this by using these legacy designs, either created by the circuit layout designers or by EDA tools, as training data to create the models.

1.4 Conclusions

The main objectives for this work are detailed below:

- Improve the automation of analog ICs by reutilizing data collected from previous circuit projects;
- Provide an overview of ML methodologies. These will then be assessed to estimate their applicability to analog IC automation;
- Create and implement functional and automated models that can learn patterns from circuit projects and can generalize that knowledge to new projects;
- Apply the developed models in analog IC projects in order to prove that the models can indeed learn reusable knowledge.

The developments of this work led to the creation of ANN models that output the device sizing, given it specifications, and a possible placement for a circuit based on the sizing of its devices. The first abstracts the details on device behavior by using the performance of previously simulated circuits to train the model. The model acts as the inverse function of the circuit simulator, given the sizes given the performance. The second abstracts the need of following complex and conflicting placement requirements by following the pattern of legacy designs that already satisfied them.

The advantage of using ANNs is that its predictions are given sizing at a push-button speed and reducing the time spent in this task of the whole IC design process. From the best of our knowledge, this is the first exploratory work in this research field that combines the potential of ANNs, to improve analog IC placement automation.

References

1. N. Lourenço, R. Martins, N. Horta, *Automatic Analog IC Sizing and Optimization Constrained with PVT Corners and Layout Effects* (Springer, 2017)
2. R. Martins, N. Lourenço, N. Horta, *Analog Integrated Circuit Design Automation—Placement, Routing and Parasitic Extraction Techniques* (Springer, Berlin, 2017)
3. G.G.E. Gielen, R.A. Rutenbar, Computer-aided design of analog and mixed-signal integrated circuits. Proc. IEEE **88**(12), 1825–1854 (2000)
4. G.G.E. Gielen, CAD tools for embedded analogue circuits in mixed-signal integrated systems on chip. IEE Proc. Comput. Digit. Tech. **152**(3), 317–332 (2005)
5. M.I. Jordan, T.M. Mitchell, Machine learning: trends, perspectives and prospects. Science **349**, 255–260 (2015)
6. C. Cadwalladr, E. Graham-Harrison, Revealed: 50 million Facebook profiles harvested for Cambridge Analytica in major data breach. The Guardian, 17 March 2018. [Online]. Available: https://www.theguardian.com/news/2018/mar/17/cambridge-analytica-facebook-influence-us-election
7. M. Bojarski, D.D. Testa, D. Dworakowski, B. Firner, End to end learning for self-driving cars (2016)
8. G. Zweig, S. Russell, Speech Recognition with dynamic bayesian networks, UC Berkeley (1998)

9. C. Bahlmann, B. Haasdonk, H. Burkhardt, *Online Handwriting Recognition with Support Vector Machines—A Kernel Approach, in Frontiers in Handwriting Recognition* (Ontario, Canada, 6–8 Aug. 2002)

10. K. Dago, R. Luthringer, R. Lengelle, G. Rinaudo, J. Matcher, Statistical decision tree: a tool for studying pharmaco-EEG effects of CNS-active drug, in Neuropsychobiology (1994), pp. 91–96

11. Cadence, Virtuoso Analog Design Environment GXL. Retrieved from http://www.cadence.com (March, 2019)

12. MunEDA, WIKED™. Retrieved from http://www.muneda.com (March 2019)

13. R. Martins, N. Lourenco, N. Horta, J. Yin, P.-I. Mak, R.P. Martins, Many-objective sizing optimization of a class-C/D VCO for ultralow-power iot and ultralow-phase-noise cellular applications. IEEE Trans. Very Large Scale Integr. Syst. **27**(1), 69–82 (2019)

14. R. Gonzalez-Echevarria et al., An automated design methodology of RF circuits by using pareto-optimal fronts of EM-simulated inductors. IEEE Trans. Comput. Des. Integr. Circ. Syst. **36**(1), 15–26 (2017)

15. G. Wolfe, R. Vemuri, Extraction and use of neural network models in automated synthesis of operational amplifiers. IEEE Trans. Comput. Aided Des. Integr. Circuits Syst. **22**(2), 198–212 (2003)

16. G. Alpaydin, S. Balkir, G. Dundar, An evolutionary approach to automatic synthesis of high-performance analog integrated circuits. IEEE Trans. Evol. Comput. **7**(3), 240–252 (2003)

Chapter 2
Related Work

2.1 Existing Approaches to Analog IC Sizing Automation

Despite the overall lack of any well-established flow for analog IC design automation, the automation of analog circuit sizing using optimization is a quite established approach [1]. These tools for automatic IC sizing vary mostly by the optimization method they use and how they estimate the circuit's performance, while in a manual design approach, the estimate can be done using approximate equations in early stages and using the circuit simulator in a later stage to verify and fine-tune the initial design.

For automatic sizing, simulation-based optimization is the most prevalent method in both industrial [2, 3] and academic [4, 5] environments, since designers prefer to avoid the risks of estimation errors in equation-based performance approximation. In fact, the trend in analog IC sizing automation is to increase performance estimation accuracy. The consideration of variability [6], models from electromagnetic simulation [7], and layout effects [8] to ensure that the optimization is done over a performance landscape that is as close as possible to the true performance of a fabricated prototype.

Nevertheless, instead of going from the target specification to the corresponding device's sizes, the designer or the electronic design automation (EDA) tool is actually going from tentative devices' sizes to the corresponding circuit performance countless times, trying combinations of design variables until a suitable sizing where the circuit meets specifications. While these EDA approaches are able to find solutions, and sometimes extremely effective ones, the long execution time while not the only is still major deterrent. More, since these approaches do not store the output of the tentative tries when a new design target appears, most of the data produced are not considered. Even the available previous designs, which sometimes are used to initialize new optimization runs, are not used to their full potential to predict effectively better starting points for the new optimization runs. Having identified this gap, and with the rise of the machine and deep learning, together with the increasing

© The Author(s), under exclusive license to Springer Nature Switzerland AG 2020   9
J. P. S. Rosa et al., *Using Artificial Neural Networks for Analog
Integrated Circuit Design Automation*, SpringerBriefs in Applied
Sciences and Technology, https://doi.org/10.1007/978-3-030-35743-6_2

availability of computational resources and data, researchers started to explore new data-centric approaches. The next section presents some of these new approaches that are capable of predicting devices sizes, given some previous designs and new target specifications or technology node.

2.1.1 Machine Learning Applied to Sizing Automation

ML, and in particular artificial neural network (ANN) models, has been used in some works to address analog IC sizing automation. In [9], ANN models for estimating the performance parameters of complementary metal-oxide semiconductor (CMOS) operational amplifier topologies were presented. In addition, effective methods for the generation of training data and consequent use for training of the model were proposed to enhance the accuracy of the ANN models. The efficiency and accuracy of the performance results were then tested in a genetic algorithm-based circuit synthesis framework. This genetic synthesis tool optimizes a fitness function based on a set of performance constraints specified by the user. Finally, the performance parameters of the synthesized circuits were validated by SPICE simulations and later compared with those predicted by the ANN models. The set of test bench circuits presented in this work can be used to extract performance parameters from other op-amp topologies other than ones specifically studied here. Circuits with different functionalities than an op-amp would need new sets of SPICE test bench circuits to create appropriate ANN models.

The ANN models trained in this work used data generated from SPICE, where time-dependent and frequency-dependent data points were created for numerous circuit topologies instantiating the target op-amp. The training set was comprised of 3095 points, while the validation set was comprised of 1020 points. The neural network toolbox from MATLAB was used to simulate the networks. The structure of the network is simply comprised of an input layer, an output layer, and a single hidden layer. Different numbers of hidden layer nodes (from 8 to 14) were iteratively tested to obtain the best possible generalization and accuracy on both training and validation examples. The hyperbolic tangent sigmoid function was used as the activation function for all hidden layer nodes, and a linear function was used as the activation function for all output layer nodes. It should be noted that each network only has one single output node since a different network was designed to model each individual op-amp performance parameter. Collecting the data took approximately 1h47 m, and generating all seven performances estimates using the ANN models took around 51.9 μs, which, when compared with using SPICE directly, resulted in a speedup factor of about 40,000 times. Still, the model is trained to replace the circuit simulator; [10] takes a similar approach.

While the improvement in the optimization efficiency is remarkable, the loss in accuracy goes against the designers' preferences preventing these, and other, performance approximation methods to establish themselves.

In [11], a multilayer perceptron (MLP), which is an ANN with a single hidden layer, was used for the migration of basic, cascode, Wilson, and regulated Wilson current mirrors between different integration technologies (1.5, 0.5, 0.35, 0.25, and 0.18 μm). The dataset was comprised of 4620, 10,632, 8480, and 9086 current mirror designs, respectively, whose performance was obtained using circuit simulation. The model was trained with all current mirror topologies, and the topology selection was binary coded as 2 inputs features of the ANN. In order to identify the technology node, one additional input was assigned to the minimum gate length. The performance figures considered were the reference current, the desired current, and the compliance voltage, leading to a total of six inputs of the ANN. The model output was the widths of the four devices (padded with zeros for the circuits with fewer devices). The authors report accuracies of 94%, but the range of the specifications is not reported. The authors also applied their technology migration approach to a differential amplifier. 100 design points were considered, 80 for training, and 20 for the test. The specifications used as inputs are the DC gain, bandwidth, input common-mode range, slew rate, and power dissipation; one additional input was the minimum channel length. The outputs are the widths for the five transistors. An accuracy of 90% considering 10% performance tolerance was reported, when testing using only the 0.18 μm designs as test data. Again, the values of the specifications for the several technologies are not reported.

In [12], a work from 2015, a method for circuit synthesis that determines the parameter values by using a set of ANNs was presented. Two different ANN architectures were considered: the MLP and the radial basis function network. Each of the two networks is optimized to output one design parameter. Hyper-parameters (such as the number of nodes from the hidden layers) from both models are tuned by a genetic algorithm. The presented methodology was tested on the design of a radio-frequency, low noise amplifier, with ten design parameters to set.

The goal of this work was to find design parameters in sequence, each one constraining the determination of the next one. The process starts with an ANN being trained to correctly determine a first design parameter, by taking a set of desired performances as input and only a single target output representing the chose design parameter. From the two architectures specified above, MLP and radial basis function network, one is chosen to characterize this ANN. This selection is performed by a genetic algorithm, which is also responsible for determining the network's hyper-parameters and which design parameter should be chosen as output. The genetic algorithm stops after an ANN implementation achieves satisfiable results during the training in terms of generalization and accuracy. The same process is repeated for a second ANN that will also take as input the output of the first ANN. This way, the second ANN will specify the second design parameter as a function of the first design parameter, as well as of the performance criteria. The process is repeated until all the design parameters have been found.

The number of points in the dataset was 235, and a varying number of them (selected by the genetic algorithm) was used for training the ANNs, while the remainder of the points were used for the test set. All the tests were performed with the MATLAB neural network toolbox. The considered ANN architectures consist of an input layer, an output layer, and a single hidden layer. The number of nodes from the hidden layer is selected by the genetic algorithm, and it can range from 2 to 62 nodes for the multilayer perceptron architecture, and from 1 to 128 nodes for the radial basis function architecture. Logarithmic sigmoid and tangent sigmoid functions were used as activation functions for the hidden layer nodes for the multilayer perceptron architecture. The method proposed in this work was able to find the ten design parameters on all twenty performed simulations. Computation times were approximately 5.375 h.

In [13], a work from 2017, a prediction method of element values of op-amp for required specifications using deep learning was proposed. The proposed method is a regression analysis model implemented via a fully connected feed-forward ANN that can learn the relation between the performances and device sizes of a target circuit. Thirteen circuit's performances were used for learning: current consumption, power consumption, DC gain, phase margin, gain–bandwidth product, slew rate, total harmonic distortion, common-mode rejection ratio, power supply rejection ratio, output voltage range, common-mode input range, output resistance, and input-referred noise. The used dataset was comprised of 13,500 different points, in which 13,490 are used as a training set and 10 are used as a test set. TensorFlow was used as the machine learning resources library. The structure of the feed-forward neural network was comprised of an input layer, two hidden layers with 100 and 200 nodes, respectively, and an output layer. Both hidden layers used rectified linear unit as an activation function. Collecting the data to feed the neural network took approximately 18 h, and only 19 min was spent on training the network. Results from simulations indicated that the proposed deep learning method succeeded in predicting the sizes for 7 devices which satisfies the required 13 performances with an accuracy of 93% on average.

As seen, ANNs are not only able to replace the circuit simulator but are also able to learn how to predict the device sizing directly. Given the target specifications and some previous designs, ANNs can be used to produce designs for specifications, not in the training set. However, in these works, it is assumed that the designer will provide target specifications matching the training/test distributions, which is not the general use case. The issues arise from the fact that analog specifications are usually defined as inequalities; e.g., when designing an amplifier, the gain is a target to be larger than some minimal value, not to be exactly defined. Nevertheless, the previously mentioned models are trying to predict sizing that meets the exact performance and therefore showing good accuracies when fed with test points, i.e., points where the exact specifications are known to have a corresponding circuit sizing. The model may fail to predict a proper design when the designer feeds it with arbitrary target specifications. This may happen even if the networks were trained and can produce designs that meet those specifications, especially if the designs meet specifications by large excess. The reason is that the ANNs trained in previous work will try to produce a design that meets exactly all performance targets (the model's inputs); however, it is likely that such design does not even exist when the target specifications are given in a real use-case scenario.

2.2 Existing Approaches to Automatic Layout Generation

Layout generation is commonly split into two different phases and therefore different problems: floorplanning and routing. Research community proposed three different methodologies for automatic analog layout synthesis, which are distinguished by the amount of legacy knowledge that is used during the generation: *layout generation considering placement and routing constraints* (i.e., without considering legacy data), *layout migration or retargeting* from a previous legacy design or template, and *layout synthesis with knowledge mining* from multiple previously designed layouts, as shown in Fig. 2.1. While none of them actually succeeded in industrial applications, recent developments in artificial intelligence and ML and, their rising success in different fields of study, indicate that the knowledge mining route might bring promising results for the automation attempts of this task. The advantages and disadvantages of each method are also highlighted in Table 2.1.

2.2.1 Layout Generation with Placement and Routing Constraints

This methodology does not use any information from previously designed layouts; instead, every solution is generated from scratch. To reduce the unpredictability of the time-consuming optimization process and produce a solution meaningful for the designer, usually, a high amount of topological constraints and objectives are set. The first step of this layout synthesis methodology consists on generating the devices required on the circuit, i.e., module instantiation, creating an independent layout of each device and/or group of devices, including the internal routing. The second step

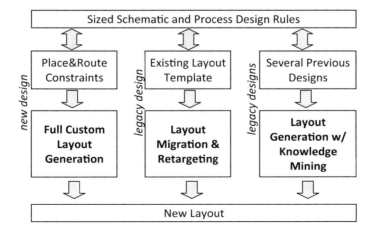

Fig. 2.1 Design flow of major analog layout synthesis methodologies [49]

Table 2.1 Advantages/disadvantages of each layout generation method

Approach	Advantages	Disadvantages
Layout generation with placement and routing constraints	No information from previously designed layouts is required	Unpredictable Time-consuming To produce a solution meaningful for the designer, a high amount of topological constraints and objectives must be set
Layout migration and retargeting	Fast generation Designers' intents from a previously designed layout are kept	A single template hardly yields compact placement solutions for the multitude of different sizing solutions that can be provided The exact same circuit topology must be available on legacy data
Layout synthesis with knowledge mining	Fast generation Combine designers' intents from different previously designed layouts	Conflicts between legacy data are deterministically solved, as the acquired "knowledge" is not generalized but only applied Parts of the circuit topology must be available on legacy data

consists of placing those devices on the floorplan while respecting the topological constraints and achieving the minimum layout area and/or desired aspect ratio. One of the most relevant factors when developing a placement tool is its representation of the cells internally, and each tool has its own strategy for it. The two main classes of approaches that have been used in the last years are distinguished by how the optimizer encodes and moves those cells [14]: by absolute representation, i.e., cells are represented by means of absolute coordinates [15–21]; and, by means of a relative representation, i.e., encoding the positioning relations between any pair of cells, the last one can further classified into slicing [22] or non-slicing representations, e.g., sequence pair [23], ordered tree [24, 25], B*-tree [26–28], transitive closure graph-based [29–32], or HB*-tree [33–35]. The placement result is commonly achieved using an optimization-based approach, being simulated annealing algorithm [36] the most common optimization kernel used. Finally, the interconnections between devices are drawn respecting the routing constraints and minimizing the total wiring length. This process is usually performed with traditional path-finding algorithms, e.g., classic maze algorithm [37] or line-expansion techniques [16]; however, fully stochastic approaches were also proposed [38–40]. The major disadvantage of this approach is that the result of it is ultimately unpredictable, and therefore, the output can be meaningful for the designer or not.

2.2.2 Layout Migration and Retargeting

Since the layouts generated using optimization and considering several constraints for placement and routing may still not exactly reproduce major designers' intents, some studies synthesize analog layouts based on layout migration or layout retargeting from legacy data. Layout retargeting is the process of generating a layout from an existing layout. The main target is to conserve most of the design choices and knowledge of the source design, while migrating it another given technology; update specifications; or attempt to optimize the old design. To synthesize the new layout with new devices' size and process technology, the previously designed layout topology is kept since it contains the designers' knowledge and the layout design preferences. Therefore, there is no need to design the entire layout from scratch. Using a qualified analog layout as the basis for a new one ensures the topology and some layout constraints are kept such as symmetry and proximity. The new layout can be generated using compaction techniques such as linear programming or graph-based algorithms that minimize the layout area and satisfy the set of placement and routing constraints [41, 42]. The migration or retargeting can also be performed using the so-called template-based approaches [8, 43]. In the latest, a technology- and specification-independent template is created by the designer, and then, that template is used for different devices' sizes and/or technologies. As major drawbacks, a single template or legacy layout hardly yields compact placement solutions for the multitude of different sizing solutions that can be provided for the same circuit, and also, the development of a placement template can be as time-consuming as manually designing the placement itself. In [44, 45], a methodology that combines *layout generation considering placement and routing constraints* and *layout migration or retargeting* was proposed, by generating placement templates through an optimization process.

2.2.3 Layout Synthesis with Knowledge Mining

More recently, a knowledge-based methodology that generates new layouts by integrating existent design expertise was proposed in [46, 47]. The approach automatically analyzes legacy design data including schematics, layouts, and constraints and generates multiple layouts for the new design by reutilizing the legacy information. In summary, layout synthesis with knowledge mining consists in four major steps: (1) analysis of the legacy design data including circuits, layouts, and constraints; (2) construction of design knowledge database resulting from the analysis of legacy designs; (3) extraction of matched subcircuits between new and legacy designs; and (4) generation of the feasible layouts for the new design by utilizing the quality-approved legacy layouts of matched subcircuits in the design knowledge database. Unlike *layout migration or retargeting* from a previous legacy design or template, here a small number of legacy layouts from different circuit topologies are used.

However, since graph representations are built for each legacy layout, and then, the solution being produced matches its subcircuits with those found on the legacy graphs, the conflicts are deterministically solved, as the acquired "knowledge" is not generalized but only applied.

2.2.4 Analog Placement Constraints

In the floorplanning task, the devices are placed in the chip area, and, in the routing task, the interconnections between those devices are drawn. The first of those problems, which is also the focus of this work, in order to reduce the unwanted impact of parasitic structures, process variations, or different operating conditions on-die, many topological requirements must be considered to produce a robust solution. It also usually involves minimizing the occupied area, wiring length, and other metrics while satisfying that huge set of symmetry, proximity, regularity, boundary, current/signal-flows, thermal, etc., constraints [14]. Furthermore, the placement quality has a definitive impact in the attainable interconnect quality, and most of the parasitic effects and consequent post-layout circuit's performance degradation are set once a placement is fixed. These facts make this task extremely hard to automate, and that is why the automatic layout generation of analog ICs has been intensively studied in the last three decades [48, 49].

According to [49], depending on the application, recent studies should consider eight placement constraints during the analog placement task:

1. *Symmetry*: Keeping symmetry in a layout reduces the effect of parasitic mismatches and the sensitivity of the circuit to process variations. Each device and/or set of devices should be symmetrically placed along a symmetry axis, and any two symmetric devices usually have the same dimensions.
2. *Symmetry Island*: Symmetric devices should have only small distance separating them in order for their electric properties to be as similar as possible [49].
3. *Common Centroid*: The devices should be decomposed into a set of equal device units, and all device units are arranged in a two-dimensional array with a common centroid. This helps minimizing systematic and random mismatch among devices whose sizes and electrical property should be the same.
4. *Proximity*: The distance between a set of devices must be restricted so that the wire length interconnecting them can be reduced as much as possible, but also, a common substrate/well can be used. Furthermore, if the devices are closer, the matching quality is increased.
5. *Regularity*: The regularity constraint states that the devices in a layout should be arranged into rows and columns making these layouts more compact and less sensitive to process variations.
6. *Boundary*: The boundary constraint restricts a device to the boundary of a rectangular area around its device group. This results in less critical net wire length leading to less parasitic interactions and a better circuit performance.

7. *Current/Signal Path*: The parasitic currents in the critical current/signal path have a great impact on the performance of a circuit. The devices in that current/signal path should be placed close to each other and most of the time in a monotonic fashion, i.e., ensuring that the current/signal flow follows a continuous path in the circuit layout, resulting in a shorter wire length and total area.

8. *Thermal*: The thermal effect may degrade the circuit performance due to the presence of thermally sensitive devices, whose electrical properties can be affected by thermal radiating devices. To mitigate this problem, it is preferred to place the radiating devices along a thermal symmetry line bisecting the circuit such that the isothermal lines are symmetric across that line. Then, the thermally sensitive devices are placed symmetrically to the radiating devices, reducing the thermal mismatch between devices.

This amount of constraints makes the decision of where a device should be located a very complex task. The methodology proposed in this book can abstract the need to follow these constraints by learning patterns from legacy designs in which they were already respected, of course, if previously considered relevant by the layout designer or EDA engineer.

2.3 Conclusions

By inspecting some successful ML applications in analog IC sizing, it was also possible to grasp more clearly which approaches might be the most desirable. Analog systems are usually characterized by a set of performance parameters that are used to quantify the properties of the circuit, i.e., design parameters. The relationship between circuit performances and design parameters can be interpreted as either a direct problem or an inverse problem. The former asserts circuit performances as a function of design parameters, while the latter deals with finding the most appropriate design parameters for a given set of performance specifications [50]. Mapping the relationship between these two features is a heavily nonlinear process, given the high interdependence among design variables. Moreover, a set of performances might be valid to more than one set of design variables; i.e., different circuit topologies might be satisfied by the same set of performances.

The problems we are trying to solve in this work are indeed inverse problems. Having this in mind, the ideal candidate solution should be able to deal with nonlinearities and try to map the complex relationship between performance figures and design parameters. This type of problem also tells us something about the nature of the model we want to build: Performances will be used as input data, and design parameters will be used as output data. Specifically, for this work, we want to obtain devices' sizes, such as the lengths and widths of transistors.

On the second part of this chapter, the state-of-the-art approaches to design analog ICs layout were overviewed. It was concluded that solutions using ML methods are yet to exist; however, most recent works [46, 47] point for an ideology of using several previously designed and verified layouts. By using legacy layouts handcrafted by layout designers or produced by EDA tools, the time of manually designing a placement, optimizing a placement solution with constraints, or designing a placement template that suits the designer's needs is bypassed. ANNs, and their capability of handling very nonlinear problems and for mapping relationships between input and output data, present themselves as a fitting choice to solve this placement issue.

References

1. N. Lourenço, R. Martins, N. Horta, Automatic Analog IC Sizing and Optimization Constrained with PVT Corners and Layout Effects (Springer, 2017)
2. Cadence, Virtuoso Analog Design Environment GXL [Online]. Available: http://www.cadence.com. Accessed 15 May 2019
3. MunEDA, WIKED™ [Online]. Available: http://www.muneda.com. Accessed am 15 May 2019
4. R. Martins, N. Lourenco, N. Horta, J. Yin, P.-I. Mak, R.P. Martins, "Many-objective sizing optimization of a class-C/D VCO for ultralow-power IoT and ultralow-phase-noise cellular applications. IEEE Trans. Very Large Scale Integr. Syst. **27**(1), 69–82 (2019)
5. F. Passos et al., A multilevel bottom-up optimization methodology for the automated synthesis of RF systems. IEEE Trans. Comput. Des. Integr. Circ. Syst. (2019). https://doi.org/10.1109/TCAD.2018.2890528
6. A. Canelas, et al., FUZYE: a fuzzy C-means analog IC yield optimization using evolutionary-based algorithms. IEEE Trans. Comput. Aided Des. Integr. Circ. Syst. (2018). https://doi.org/10.1109/tcad.2018.2883978
7. R. Gonzalez-Echevarria et al., An Automated design methodology of RF circuits by using pareto-optimal fronts of EM-simulated inductors. IEEE Trans. Comput. Des. Integr. Circ. Syst. **36**(1), 15–26 (2017)
8. R. Martins, N. Lourenço, F. Passos, R. Póvoa, A. Canelas, E. Roca, R. Castro-López, J. Sieiro, F.V. Fernandez, N. Horta, Two-step RF IC block synthesis with pre-optimized inductors and full layout generation in-the-loop. IEEE Trans. Comput. Des. Integr. Circ. Syst. (2018)
9. G. Wolfe, R. Vemuri, Extraction and use of neural network models in automated synthesis of operational amplifiers. IEEE Trans. Comput. Aided Des. Integr. Circ. Syst. **22**(2), 198–212 (2003)
10. H. Liu, A. Singhee, R. A. Rutenbar, L.R. Carley, Remembrance of circuits past: macromodeling by data mining in large analog design spaces, in *Proceedings 2002 Design Automation Conference* (2002) . https://doi.org/10.1109/dac.2002.1012665
11. N. Kahraman, T. Yildirim, Technology independent circuit sizing for fundamental analog circuits using artificial neural networks, in *Ph.D. Research in Microelectronics and Electronics* (2008)
12. E. Dumesnil, F. Nabki, M. Boukadoum, RF-LNA circuit synthesis using an array of artificial neural networks with constrained inputs. in *Proceeding IEEE International Symposium on Circuits and Systems*, vol. 2015-July (2015), pp. 573–576
13. N. Takai, M. Fukuda, Prediction of element values of OPAmp for required specifications utilizing deep learning, in *International Symposium on Electronics and Smart Devices (ISESD)* (2017)
14. R. Martins, N. Lourenço, N. Horta, Analog Integrated Circuit Design Automation—Placement, in *Routing and Parasitic Extraction Techniques* (Springer, 2017). ISBN 978-3-319-34060-9
15. D. Jepsen, C. Gelatt, Macro placement by Monte Carlo annealing, in *IEEE International Conference on Computer Design (ICCD)* (1983), pp. 495–498

16. J. Cohn, J. Garrod, R.A. Rutenbar, L. Carley, KOAN/ANAGRAM II: new tools for device-level analog placement and routing. IEEE J. Solid-State Circ. (JSSC) **26**(3), 330–342 (1991)
17. K. Lampaert, G. Gielen, W. Sansen, A performance-driven placement tool for analog integrated circuits. IEEE J. Solid-State Circ **30**(7), 773–780 (1995)
18. E. Malavasi, E. Charbon, E. Felt, A. Sangiovanni-Vincentelli, Automation of IC layout with analog constraints. IEEE Trans. Comput. Aided Des. Integr. Circ. Syst. (TCAD) **15**(8), 923–942 (1996)
19. R. Martins, N. Lourenço, N. Horta, Analog IC placement using absolute coordinates and a hierarchical combination of Pareto optimal fronts, in *2015 11th Conference on Ph. D. Research in Microelectronics and Electronics* (2015)
20. R. Martins, R. Póvoa, N. Lourenço, N. Horta, Current-flow & current-density-aware multi-objective optimization of analog IC placement. Integr., VLSI J. (2016). https://doi.org/10. 1016/j.vlsi.2016.05.008
21. R. Martins, N. Lourenço, R. Póvoa, N. Horta, On the exploration of design tradeoffs in analog ic placement with layout-dependent effects, in *International Conference on SMACD* (Lausanne, Switzerland, 2019)
22. D.F. Wong, C.L. Liu, A new algorithm for floorplan design, in *Proceedings of the 23th ACM/IEEE Design Automation Conference (DAC)* (1986), pp. 101–107
23. H. Murata, K. Fujiyoshi, S. Nakatake, Kajitani. VLSI module placement based on rectangle-packing by the sequence-pair, in *IEEE Transactions on Computer-Aided Design of Integrated Circuits and Systems*, vol. 15, no. 12 (1996), pp. 1518–1524
24. Y. Pang, F. Balasa, K. Lampaert, C.-K. Cheng, Block placement with symmetry constraints based on the o-tree non-slicing representation, in *Proceedings ACM/IEEE Design Automation Conference* (2000), pp. 464–467
25. P.-N. Guo, C.-K. Cheng, T. Yoshimura, An O-tree representation of nonslicing floorplan and its applications, in *Proceedings of the 36th ACM/IEEE Design Automation Conference (DAC)* (1999), pp. 268–273
26. F. Balasa, S.C. Maruvada, K. Krishnamoorthy, On the exploration of the solution space in analog placement with symmetry constraints. IEEE Trans. Comput. Aided Des. Integr. Circ. Syst. (TCAD) **23**(2), 177–191 (2004)
27. Y.-C. Chang, Y.-W. Chang, G.-M. Wu, S.-W. Wu, "B*-trees: A new representation for nonslicing floorplans, in *Proceedings of the 37th ACM/IEEE Design Automation Conference (DAC)* (2000), pp. 458–463
28. F. Balasa, S.C. Maruvada, K. Krishnamoorthy, Using red-black interval trees in device-level analog placement with symmetry constraints, in *Proceedings of the Asian and South Pacific—Design Automation Conference (ASP-DAC)* (2003), pp. 777–782
29. L. Jai-Ming, C. Yao-Wen, TCG: a transitive closure graph-based representation for non-slicing floorplans, in *Proceedings of the 38th ACM/IEEE Design Automation Conference (DAC)*, pp. 764–769 (2001)
30. L. Lin, Y.-W. Chang, TCG-S orthogonal coupling of P-admissible representations for general floorplans. IEEE Trans. Comput. Aided Des. Integr. Circ. Syst. (TCAD) **23**(5), 968–980 (2004)
31. L. Zhang, C.-J. Shi, Y. Jiang, Symmetry-aware placement with transitive closure graphs for analog layout design, in *Proceeding IEEE/ACM Asia and South Pacific Design Automation Conference* (2008), pp. 180–185
32. J.-M. Lin, G.-M. Wu, Y.-W. Chang, J.-H. Chuang, placement with symmetry constraints for analog layout design using TCG-S, in *Proceeding IEEE/ACM Asia and South Pacific Design Automation Conference*, vol. 2 (2005), pp. 1135–1138
33. P.-H. Lin, Y.-W. Chang, S.-C. Lin, Analog placement based on symmetry-island formulation. IEEE Trans. Comput. Aided Des. (TCAD) **28**(6), 791–804 (2009)
34. P.-H. Lin, S.-C. Lin, Analog placement based on novel symmetry-island formulation, in *Proceedings of the 44th ACM/IEEE Design Automation Conference (DAC)* (2007), pp. 465–470
35. P.-H. Lin, S.-C. Lin, Analog placement based on hierarchical module clustering, in *Proceedings of the 45th ACM/IEEE Design Automation Conference (DAC)* (2008), pp. 50–55

36. B. Suman, P. Kumar, A survey of simulated annealing as a tool for single and multiobjective optimization. J. Oper. Res. Soc. **57**, 1143–1160 (2006)
37. Y. Yilmaz, G. Dundar, Analog Layout Generator for CMOS Circuits. IEEE Trans. Comput. Aided Des. Integr. Circ. Syst. (TCAD) **28**(1), 32–45 (2009)
38. R. Martins, N. Lourenço, A. Canelas, N. Horta, Electromigration-aware and IR-Drop avoidance routing in analog multiport terminal structures, in *2014 Design, Automation & Test in Europe Conference & Exhibition (DATE)* (2014)
39. R. Martins, N. Lourenco, N. Horta, Routing analog ICs using a multi-objective multi-constraint evolutionary approachAnalog. Integr. Circ. Signal Process. **78**(1), 123–135 (2013)
40. R. Martins, N. Lourenço, N. Horta, Multi-objective multi-constraint routing of analog ICs using a modified NSGA-II approach, in *International Conference on Synthesis, Modeling, Analysis and Simulation Methods and Applications to Circuit Design (SMACD)* (Seville, Spain, 2012), pp. 65–68
41. N. Jangkrajarng, S. Bhattacharya, R. Hartono, C. Shi, IPRAIL—Intellectual property reuse-based analog IC layout automation. Integr. VLSI J. **36**(4), 237–262 (2003)
42. S. Bhattacharya, N. Jangkrajarng, C. Shi, Multilevel symmetry-constraint generation for retargeting large analog layouts. IEEE Trans. Comput. Aided Des. Integr. Circ. Syst. (TCAD) **25**(6), 945–960 (2006)
43. R. Martins, N. Lourenço, N. Horta, LAYGEN II—automatic analog ICs layout generator based on a template approach, in *Genetic and Evolutionary Computation Conference (GECCO)* (Philadelphia, USA, 2012)
44. R. Martins, A. Canelas, N. Lourenço, N. Horta, On-the-fly exploration of placement templates for analog IC layout-aware sizing methodologies, in *2016 13th International Conference on Synthesis, Modeling, Analysis and Simulation Methods and Applications to Circuit Design (SMACD)* (2016), pp. 1–4
45. R. Martins, N. Lourenço, A. Canelas, N. Horta, Stochastic-based placement template generator for analog IC layout-aware synthesis. Integr. VLSI J. **58**, 485–495 (2017)
46. P.H. Wu, M. P. H. Lin, T.Y. Ho, Analog layout synthesis with knowledge mining, in *2015 European Conference on Circuit Theory and Design (ECCTD)* (2015), pp. 1–4
47. P.H. Wu, M.P.H. Lin, T.C. Chen, C.F. Yeh, X. Li, T.Y. Ho, A novel analog physical synthesis methodology integrating existent design expertise. IEEE Trans. Comput. Aided Des. Integr. Circ. Syst. **34**(2), 199–212 (2015)
48. H.E. Graeb (ed.), *Analog Layout Synthesis: A Survey of Topological Approaches* (Springer, 2011)
49. M.P.-H. Lin, Y.-W. Chang, C.-M. Hung, Recent research development and new challenges in analog layout synthesis, in *Asia and South Pacific Design Automation Conference* (2016), pp. 617–622
50. M.V. Korovkin, V.L. Chechurin, M. Hayakawa, *Inverse Problems in Electric Circuits and Electromagnetics* (Springer, 2007)

Chapter 3
Overview of Artificial Neural Networks

3.1 Machine Learning Overview

The possibility of assigning tasks to machines to avoid repetitive actions as a theoretical formulation has been postulated through the ages, and it is recurring in our collective imagination. Nowadays, artificial intelligence (AI) has found application in many areas including automated customer support and tailoring, advanced driver assistance systems, etc. ML borrows its foundations from the concepts of early AI research, and its approach to problem-solving is focused on statistical knowledge. This dissonance caused ML to branch out from AI and reorganize itself as a separate field of research, since AI researchers were more concerned with automatically constructing expert systems modeled using heavily symbolic language. ML researchers favor models from statistics and probability theory [1].

Thomas Bayes laid the foundations for statistical learning on his essay on probability theory (1763) [2]. The proposed key concepts, such as conditional probability, would later consolidate in what is now called Bayes' theorem and would be of immense importance in the formulation of some early ML techniques, such as Naive Bayes or Markov chains [3]. The early movement in ML was also characterized by an emphasis on symbolic representations of learned knowledge, such as production rules, decision trees, and logical formulae [1]. The research continued and other discoveries were made, with the invention of the first neural network machine (1951) being one of the most important. However, it was not until Frank Rosenblatt invented the perceptron (1957) [4], a classification algorithm that makes its predictions based on a linear predictor function combining a set of weights with the feature vector that ANN began to receive more attention from other researchers. In 1986, the backpropagation process was proposed, and it launched even further ANNs development [5], and in the 1990s ANNs and support vector machines (SVMs) became widely popular, as the available computational power started to increase. Since then, many other discoveries have been done [6], and today, ML is a large field that involves many techniques [7, 8]. Enabling ML's success with virtual personal assistants, mail spam

J. P. S. Rosa et al., *Using Artificial Neural Networks for Analog Integrated Circuit Design Automation*, SpringerBriefs in Applied Sciences and Technology, https://doi.org/10.1007/978-3-030-35743-6_3

filters, product recommendations, or fraud detection applications used by billions of people around the world every day.

Despite its practical and commercial successes, ML remains a young field with many underexplored research opportunities. Some of these opportunities can be seen by contrasting current ML approaches to the types of learning we observe in naturally occurring systems such as humans and other animals, organizations, economies, and biological evolution. For example, whereas most ML algorithms are targeted to learn one specific function or data model from one single data source, humans clearly learn many different skills and types of knowledge, from years of diverse training experience in a simple-to-more-difficult sequence (e.g., learning to crawl, then walk, then run) [6].

ML systems can be classified in broad categories based on:

- Whether or not they are trained with human supervision (supervised, unsupervised, semi-supervised, and reinforcement learning);
- Whether or not they can learn with new data on the fly (online vs. batch learning);
- Whether they only compare new data points to already known data points or if they learn patterns present in the data (instance-based vs. model-based training).

These classification criteria of ML systems are not exclusive; i.e., they can be combined and ML techniques can be used differently in different systems. For example, ANNs can be a batch, model-based, supervised learning system but the models can also learn online with reinforcement.

3.1.1 Supervised or Unsupervised Learning

The amount and type of supervision a system gets during training define its type according to these criteria.

In **supervised learning**, the data that are fed to the system include the desired solution, called label. A classification problem is a typical example of this type of system since the algorithm is given data points that are classified with the correct label and it learns the statistical properties or patterns in those data points to achieve said label. A regression problem can also be solved using a supervised learning system, since the objective is to predict a numeric value, given a set of features (input variables). To train these systems, one must feed the algorithm with data points which each contains a set of features and the numeric value it must predict. The system then learns the relation between the output and the set of features. Some of the more important supervised learning algorithms are linear regression, logistic regression, polynomial regression, decision trees, SVMs (Fig. 3.1a), and ANNs.

In **unsupervised learning**, the data used to train the system are unlabeled. These algorithms form groups of data points based on their features without attributing a class to each group since there are no labels. This is the main focus of clustering and visualization algorithms. Dimensionality reduction is a related task to these systems, and its objective is to simplify the data while keeping the information by merging

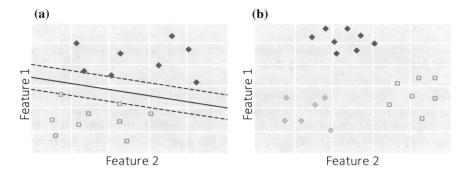

Fig. 3.1 **a** An example of a supervised learning algorithm, where a support vector machines technique was applied to classify the data into two different classes; **b** an example of an unsupervised learning algorithm, where a k-means technique was used to classify data into three different classes

features that are strongly correlated. This method is called feature extraction, and it is used to reduce the input dimensions and make it easier to visualize the grouped data points. These systems are useful for anomaly detection since it groups the "normal" data points. If a new data point stays far away from the "normal" group, it might be an anomaly. Some of the more important unsupervised learning algorithms are clustering algorithms like k-means and visualization and dimensionality reduction algorithms like principal component analysis (Fig. 3.1b).

In **semi-supervised learning**, the data used to train the system are partially labeled. Usually, the training data have a lot of unlabeled data points and a few labeled ones. Most semi-supervised learning algorithms are combinations of supervised and unsupervised learning algorithms. For example, deep belief networks are based on unsupervised components called restricted Boltzmann machines that are trained in an unsupervised manner and then the whole system is fine-tuned using supervised learning techniques.

Reinforcement learning implies the existence of an agent that can observe the environment and interact with it by selecting and executing actions. The agent gets a reward based on the action it chose (the reward can be positive or negative). It must learn what is the best strategy, called policy, to maximize the reward over time. A policy is a function that defines what action the agent should choose when it is in a given situation. These systems are used in robots to teach them how to walk. Reinforcement learning was also used on DeepMind's AlphaGo program that beat the Go game world champion [9]. It learned its policies by analyzing millions of games and then playing against itself.

3.1.2 Batch or Online Learning

This classification distinguishes systems based on their capacity to learn incrementally from a stream of incoming data.

In **batch learning**, the system is trained using all the available data. This process takes a lot of time and requires demanding computational resources. For this reason, the training is usually done before the system is used to predict the output for new data: First, the system is trained, and then, it is used. This is called offline learning. To train an already trained batch learning system with new data, it might be necessary to train a new version of the system from scratch with the new dataset (old data augmented with the new data) and then replace the old system with the newly trained one. As this process can take hours to execute, it is usually done weekly on commercial systems. The upside is that it is easy to automate the training, evaluation, and launching of a new system. Another of the limitations of these systems is that if the datasets are large, there might not be enough computational resources available to train with the full dataset, making it impossible to use a batch learning algorithm.

In **online learning**, the system is trained incrementally by feeding new data sequentially, either individually or in small groups called mini-batches. Contrarily to what happens in batch learning, these training steps are fast and cheap in terms of computational resources, so it is possible for the system to learn about new data points while it is running. These systems are perfect when new data is available at a continuous flow or when the computational resources are limited. The downside is the fact that these systems are always consuming and learning about new data, so if bad/wrong data is fed to the system, its performance will start to degrade.

3.1.3 Instance-Based or Model-Based Learning

This criterion evaluates the approach the systems take to generalize to new data. Almost every ML task has, as its objective, made predictions, so even if it has a 100% accuracy on the training set, it has to generalize to examples it has never seen before.

Instance-based learning systems use a similarity measure to generalize to new data. This similarity measure is used to compare the new data to the already known data and output a predicament based on that value. Clustering algorithms are an example of such systems since new data is attributed to an existing cluster based on the distances to those clusters.

Model-Based learning systems, as the name suggests, build a model from the training data to generalize to new data. Usually, models are defined by a set of parameters that have to be learned during the training step. To measure what set of parameters achieve a better performance, a cost (utility) function is defined that evaluates how bad (good) the model is. ANNs are the prime example of this kind of system since they define a model by themselves.

3.2 Challenges in Creating ML Systems

An ML system is a learning algorithm trained on data. Therefore, there are two things that can go wrong when designing such a system: A problem on the algorithm or a problem in the data used to train it. Most ML algorithms need a large amount of data to work properly, and even for simple problems, the dataset should contain thousands of examples, while for more complex programs such as image recognition there should be millions of examples. Provided there is enough data, different ML algorithms can perform identically well on a complex problem. This was shown by Banko and Brill in [10], where different algorithms had a similarly good performance in a problem of natural language disambiguation. These results suggest that might be more important to spend time and money on acquiring more and better data then on the development of the algorithm. However, small- and medium-sized datasets are still very common and it is not easy nor cheap to acquire more data.

In order to generalize well, the algorithm has to have access to all cases it has to generalize to at training time; i.e., the training data must contain examples of every case the algorithm needs to learn about. If this is not the case, the system might start to have low accuracy when it starts to be presented with inputs not represented by the training data. In addition, the training data can contain errors, outliers, and noise. This makes it harder for the system to learn patterns in the data, which will impact its performance. It is often worth to clean the data before feeding it into a learning algorithm.

Furthermore, the system can only learn if the data contain enough relevant features and a few irrelevant ones. Thus, it is important to do a process called feature engineering that involves selecting the most useful features, combining existing features into a single, more useful one (by using dimensionality reduction algorithms, for example) and creating new features by acquiring new data or extending the available data (e.g., polynomial features, images shifting, etc.). This will ensure that the system trains with useful data.

Preventing overfitting is a major concern when designing an ML system. An overfitting model performs well on the training data but does not generalize well to new data points, which is the main objective of an ML system: Predict a correct output for data that it has not seen before. Complex models are more prone to overfit since they can more easily detect patterns within the data. However, if there are not too many examples or if they are noisy, the model can start to detect patterns in the noise itself. Overfitting usually happens when the model is too complex for the amount and/or noisiness of the training data. To prevent overfitting, one can simplify the model by selecting one with fewer parameters, by reducing the number of attributes in the training data or by constraining the model. Obviously, increasing the number of examples or reducing the amount of noise (by removing outliers and fixing errors) in the training data also helps in the prevention of overfitting.

Contrary to overfitting, underfitting can also be a problem in an ML system. This can happen when the model is too simple, and so, it fails to learn the patterns in the data. This can be prevented by making the model more complex by increasing the number of parameters or by feeding better features to the learning.

3.3   The Five Tribes of Machine Learning

While the methods presented in this book mostly use deep learning and ANNs, the following section provides a brief overview of the broader range of ML algorithms, as many of them can also provide alternatives for electronic design automation (EDA). ANNs are a good (but not the only) choice for analog integrated circuit (IC) design automation. The following exposition follows the taxonomy proposed by Domingos in his book, *The Master Algorithm* [11], that separates ML algorithms in five different "tribes" based on their core methods and philosophy. Those tribes are the symbolists, Bayesians, connectionists, evolutionaries, and analogizers. Table 3.1 shows a summary of the advantages and disadvantages of each tribe.

All five tribes have something unique to offer. The key is to understand what type of data is available and figure out which methodology will best suit the problem at hand. Different problems require different approaches, and the next sections go over the strengths and weaknesses of several ML approaches and analyze their relevance in the context of electronic design automation. Some problems may be solved using only one technique, while others might require a combination of different techniques. The size and structure of the data are important properties to analyze early. Assessing the type of learning problem (supervised or unsupervised) being handled is also an easy way to exclude some options. Computation time is also an important property to take into account.

Symbolists argue that knowledge can be reduced to manipulating symbols, the same way a mathematician solves equations by replacing expressions by other expressions. Instead of starting with an initial premise and looking for a conclusion, the inverse deduction starts with some premises and conclusions and essentially works backward to fill in the gaps. This is done by deducing missing rules that fit the pre-established conclusions (much like solving a puzzle). Representative algorithms are rules and decision trees.

Symbolists are probably the simplest to understand, interpret, and visualize set of techniques one can come across. Rules and decision tree offer clear insight about data and are easy to extrapolate conclusions from. They can handle both numerical and categorical data and multi-output problems. Nonlinearities in the data do not affect the performance of trees. Nevertheless, complex datasets where several rules need to be inferred may not be the most fitting for these techniques. Overcomplex trees may end up not generalizing the data well, thus resulting in overfitting. Some classes may also dominate others if the input dataset is not balanced, so it is recommended to have a similar number of examples for each class (this is true for almost all algorithms). In terms of achieving satisfying results, a global maximum may not be found by simply running one decision tree. A more favorable set of results can be obtained by training multiple trees where features and samples are randomly sampled with

Table 3.1 Advantages and disadvantages between each tribe of ML

Tribe	Representative method	Advantages	Disadvantages
Symbolists	Rules and decision trees	Easy to understand; Can handle both numerical and categorical data and multi-output problems	Not a practical approach if there are several decisions to be made Overcomplex trees may end up not generalizing the data well, thus resulting in overfitting
Bayesians	Naïve Bayes and Markov chains	Requires fewer data to be effective Fast in the prediction of a given feature's class Insensitive to irrelevant features	Not useful when approaching real-life situations, since features are usually interdependent
Connectionists	Artificial neural networks	Appropriate for complex and noisy datasets Good generalization of rules between input features and output values.	Might be necessary to iterate the algorithm several times to yield favorable results Overfitting is likely to occur
Evolutionaries	Evolutionary algorithms and genetic programming	Appropriate for problems with a wide range of parameters	Expensive computation times May not find global maxima
Analogizers	Support vector machines	Can solve both linear (classification) and nonlinear (regression) problems It can produce unique solutions	Inappropriate for complex and noisy datasets Undesirable choice when tackling a problem with high-dimensional data

replacement. Random forests generalize this concept and are used to overcome the overfitting problem inherent to decision trees. Both training and prediction are very fast, because of the simplicity of the underlying decision trees. In addition, both tasks can be straightforwardly parallelized, because the individual trees are entirely independent entities.

Bayesians are concerned above all with uncertainty. This type of learning evaluates how likely a hypothesis will turn out to be true while considering a priori knowledge. Different hypotheses are compared by assessing which outcomes are more likely to happen. This is called probabilistic inference. This tribe's favored algorithms are Naïve Bayes or Markov Chains.

The Naive Bayes, from the **Bayesians** Tribe, algorithm affords fast, highly scalable model building and scoring. Naive Bayes can be used for both binary and multiclass classification problems. These models are relatively easy to understand and build. They are easily trained and do not require big datasets to produce effective results. They are also insensitive to irrelevant features. These advantages mean that a Naive Bayesian classifier is often a good choice for an initial baseline classification. If it performs suitably, it means you will have a very fast and interpretable classifier for your problem without much effort. If it does not perform well, other models should be explored. Since this algorithm always assumes that features are independent, which is not true for most real-life situations, this model will not perform well in many cases.

For **analogizers**, the key to learning is recognizing similarities between situations and thereby inferring other similarities. Learning comes down to building analogies between available data. The tribe of the **analogizers** is better known for the SVM technique. It is one of the most efficient ML algorithms and is mostly used for pattern recognition [12]. It has a wide range of applications, such as speech recognition, face detection, and image recognition. This is a very powerful supervised learning algorithm for data separation which builds a model that foresees the category of a new example, based on a given set of featured examples. It works on the principle of fitting a boundary to a region of points that meet the same criteria of classification, i.e., belong to the same class. Once a boundary is fitted on the training sample points, for any new points that need to be classified, the designer must only check whether they are inside the boundary or not. The advantage of SVM is that once a boundary is established, most of the training data is redundant. All it needs is a core set of points that can help identify and set the boundary. These data points are called support vectors because they "support" the boundary. Data types for this algorithm include linear and nonlinear patterns. Linear patterns are easily distinguishable and can be separated in low dimensions, but the same cannot be said about nonlinear patterns. The latter needs to be manipulated for the data to become separable, e.g., by means of kernel functions. Another advantage of SVMs is that they generalize new samples very well. When an optimal set of hyperplanes that separate the data is achieved, SVMs can produce unique solutions, which is a fundamental difference between this technique and ANNs. The latter yields multiple solutions based on local minima, which might not be accurate over different test data. In terms of disadvantages, SVMs might not be the most desirable choice when tackling problems with high-dimensional data. And, they are heavily reliant on the choice of the kernel and its parameters, and even then, the obtained results might not be transparent enough to extrapolate any meaningful conclusions.

Evolutionaries believe that the mother of all learning is natural selection. In essence, an evolutionary algorithm is a meta-heuristic optimization algorithm used in AI that uses mechanisms based on biological evolution. Beyond their ability to search large spaces for multiple solutions, these algorithms are able to maintain a diverse population of solutions and exploit similarities of solutions by mechanisms of recombination, mutation, reproduction, and selection [13]. A fitness function is responsible for analyzing the quality of the proposed solutions, outputting values that will then be compared to a predefined cost or objective function, or a set of optimal trade-off values in the case of two or more conflicting objectives.

Evolutionary algorithms, from the **evolutionaries** tribe, are a set of modern heuristics used successfully in many applications with great complexity. The most immediate advantage of evolutionary computation is that it is conceptually simple. Almost any problem that can be formulated as a function optimization task can be modeled by an evolutionary algorithm [14]. These algorithms are capable of self-optimization, and their processes are highly parallel. This means that the evaluation of each obtained solution can be handled in parallel. Only the mechanism of selection requires some serial processing. Evolutionary algorithms can also be used to adapt solutions to changing environments. Usually, it is not necessary to obtain new populations at random and restart the model when new solutions want to be obtained. This is due to the high flexibility of the algorithm, making the available populations a solid foundation to make further improvements. The disadvantages of this algorithm include the lack of guarantee of finding global maxima and high computation times. Usually, a decent-sized population and generations are needed before good results are obtained.

For **connectionists**, learning is what the brain does, and so the goal is to reverse engineer it. Models belonging to this tribe attempt to build complex networks comprised of nodes that resemble neurons from the brain, and adjust the strength of each connection by comparing obtained outputs with desired ones. Their favored algorithm is backpropagation, a method commonly applied to ANNs.

Connectionists offer a wide range of applications. Deep learning has become quite popular in the last few years in image processing, speech recognition, and other areas where a high volume of data is available. ANNs, the preferred model of this tribe, can yield some impressive results since they can generalize rules between input features and output variables, given we have enough examples in our dataset. This technique is well suited to problems in which the training data correspond to noisy, complex sensor data, such as inputs from cameras and microphones. The target function output may be discrete-valued, real-valued, or a vector of several real- or discrete-valued attributes. It is an appropriate technique for datasets that contain training examples with errors, and while ANN learning can take relatively long, evaluating the learned network is typically very fast. However powerful these techniques may be, achieving the most favorable model architecture is a challenging and iterative process. There are several details the designer needs to be mindful of when crafting an ANN to achieve a low training error of the network while avoiding overfitting, namely the number of layers, activation functions, the loss function and optimizers used or the normalization of the data. Later in this chapter, a detailed review of the model choices is presented.

3.3.1 Why Use ANNs for EDA

All these approaches provide valuable techniques that can be applied in analog IC design automation. From all the above-mentioned methods, evolutionary is probably the most used approach in analog EDA, namely for global optimization of analog IC sizing. NSGA-II is at the core of many state-of-the-art EDA optimization approaches [15–18].

SVMs can provide high-accuracy classifiers and have well-documented methods to avoid overfitting, and with an appropriate kernel, they can work well even if the training data are not linearly separable in the base feature space. SVMs are not as widespread in analog EDA, but they can be found in a couple of works. For example, in [19] and [20] they are used in an evolutionary optimization loop to classify if the tentative solutions are worth (or not) to be passed to the electrical stimulator. Markov chains, probabilistic graphical models, or Naive Bayes classifiers are not common in analog EDA, but a couple of recent works do employ Bayesian model fusion [21] and Bayesian optimization in analog EDA [22].

Decision trees are easy-to-interpret algorithms. They easily handle feature interactions and they are non-parametric, so the designer does not have to worry about outliers or whether the data are linearly separable. One disadvantage is that they easily overfit, but a workaround to this problem is to use ensemble methods like random forests. These methods are fast and scalable and require a low amount of tuning. Their applicability in the context of EDA is still an open topic.

This work focuses on ANNs that are widely used in pattern recognition because of their ability to generalize and to respond to unexpected inputs/patterns. ANNs are a low bias/high variance classifiers appropriate for high-volume datasets. These are flexible models that allow the implementation of different tasks in the same networks. ANNs can share features, and they can easily be trained with flexible cost functions. A complete overview of ANNs can be found in [23]. In the context of analog EDA, ANNs are an appropriate choice because their flexibility allows the application to several tasks, while the scalability might enable end-to-end models. More, the available computational power is enough to produce the large amount of data needed to train effective models.

3.4   Neural Network Overview

The building block of ANNS is the artificial neuron (also referred to as node). The operations that each neuron performs are simple: It sums its weighted inputs, passes that results through a function called activation function (usually nonlinear) and outputs that value, as shown in Fig. 3.2. Although these units execute simple operations,

Fig. 3.2 Representation of an artificial neuron

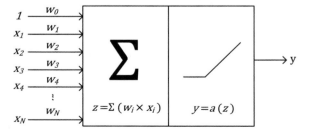

Fig. 3.3 Fully connected
ANN with two hidden layers

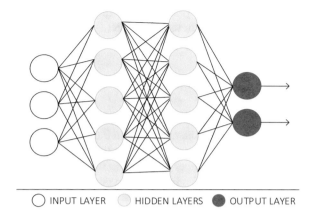

when they are grouped in dense networks, they can represent the most complex functions [24]. The parameters that define these functions are the weights of each neuron and are determined during training.

The fundamental structure for an artificial neural network is the multilayer perceptron (MLP) shown in Fig. 3.3. It consists of three or more layers of nodes (or neurons): One input layer, one output layer, and one or more hidden layers. The input layer is only responsible for distributing the inputs to the next layer: It does not do any operation with them. The number of neurons in the input layer is always equal to the number of inputs of the model. The neurons in the hidden layers accept the outputs of the neurons of the previous layer and output them to the next layer. These layers can have any number of neurons. The output layer receives the output of the last hidden layer, and its output corresponds to the output of the model. It is also possible to design a network with only an input and an output layer (without hidden layers). The most common layer structure in earlier models would have each neuron's output as an input to every neuron in the next layer like it is represented in the ANN of Fig. 3.3. These layers are called fully connected layers, other configurations are possible, such as convolutional layers where the nodes' output is connected only to some nodes in the next layer [25], or residual layers where the outputs of some layer are passed over the next layer and added before the activation function of the following layer [26].

The MLP is not the only type of ANN, and different models such as convolutional neural networks (CNNs) and recurrent neural networks (RNNs) were shown to perform well for specific types of problems. For example, CNNs are extremely well suited for image processing problems, and in addition to fully connected layers, they have convolutional layers, built with nodes that convolve filters to the input matrix for feature extraction. The filter weights are determined during training. RNNs structured in such a way that allows the processing of variable-length sequences. They are well suited for natural language processing (NLP) problems like handwriting recognition and speech recognition because part of the output at time t is passed as input for time t + 1 (and t-1 in bidirectional networks) which represents a way of memorizing, which in the case of NLP problems introduces a sense of context.

3.4.1 Size of the Model

The hyper-parameters that define the number of nodes of an ANN are the number of layers (or depth) of the network and the number of nodes in each layer (or width). In addition, the structure of the connection (fully connected, shared weights, etc.) also affects the total number of parameters to be fitted during the train. The number of nodes in an ANN is an important hyper-parameter and defines the complexity of the function that the ANN can represent. In general, a bigger number of layers (deeper model) will result in an ANN that can learn more complex functions given the same number of nodes [27], but it also increases the complexity of training the network. Deep networks are difficult to train due to numeric instability; namely, issues like the vanishing or exploding gradient can greatly impair the ability to update the weights effectively.

Regarding the number of neurons, it is simple to set the number of units in the input and output layer as they are the number of input and output variables, respectively. However, the hidden layers cause a problem where too many neurons can represent more complex functions but have a negative impact on the training time of the model since the number of weights to be computed increases greatly. There is no objective way to determine the optimal number of neurons in each hidden layer of an ANN. Some heuristics lead to good results in certain problems, but they do not generalize well, making the process of choosing the number of neurons in the hidden layers ultimately a trial-and-error process.

To have a starting point on the number of neurons (N_h for hidden layer h), some of the heuristics that can be experimented with are:

- For two hidden layer networks, the formulas $N_{h=1} = \sqrt{(m + 2)N} + 2\sqrt{\frac{N}{m+2}}$ and $N_{h=2} = m\sqrt{\frac{N}{m+2}}$, where N is the number of samples to be learned with low error and m is the number of output variables, are reported to lead to a network that can learn N samples (input–output pairs) with a small error [28].

- In [29], the formula $N_{h=L} = \frac{N_{in} + \sqrt{N_p}}{L}$ was tested for 40 different test cases. From the obtained results, the authors stated that it can be used to compute the optimum number of neurons in the hidden layers. In this formula, N_{in} is the number of input neurons, N_p is the number of input samples, and L is the number of the hidden layer.

- [30] proposes a theorem to compute the necessary number of hidden units to make the ANN achieve a given approximation order in relation to the function it is trying to model. The approximation order N implies that the derivative of order N at the origin is the same for the function defined by the ANN, $g(x)$ and for the function it is trying to model, $f(x)$, or: $g^{(N)}(0) = f^{(N)}(0)$. To achieve an approximation order

N, the number of hidden units across all hidden layers has to follow condition (3.1), where n_0 is the number of inputs of the model.

$$\sum_{h=1}^{L} N_h \geq \frac{\binom{N+n_0}{n_0}}{n_0+2}$$

when

$$\binom{N+n_0}{n_0} \leq n_0^2 + 3n_0 + 2\sqrt{n_0^3} + 4\sqrt{n_0} \qquad (3.1)$$

otherwise $\displaystyle\sum_{h=1}^{L} N_h \geq 2\sqrt{\binom{N+n_0}{n_0} + 2n_0 + 2} - n_0 - 3$

- There is some rule of thumb that can be used to choose the total number of hidden neurons, such as:

  - The number of hidden neurons should be between the size of the input layer and the size of the output layer;
  - The number of hidden neurons should be 2/3 of the size of the input layer plus the size of the output layer;
  - The number of hidden neurons should be less than twice the size of the input layer.

- One of the simplest rules is trial and error. It works by training networks with different combinations of the number of units for the hidden layers and saves the one that brings better results. The problem with this method is that it is time-consuming since training an ANN usually takes a long time. Initially, the number of hidden neurons can be set using one of the heuristics above but in the end, the fine-tuning of these parameters should be adjusted using this method in search of the model that leads to better results.

As seen, setting an appropriate depth and number of nodes is of the utmost importance, as it defines the models' expressiveness. However, it is difficult to guess the right number of layers and the number nodes in each layer that make for the best model. Hence, in order to find the best shape of an ANN to solve the desired problem, a good first step is to train a network with two layers, which corresponds to a simple multivariate linear model. The evaluation of this model allows us to estimate how difficult are the data and provides a baseline for improvement, then increasing the size of the model (width and depth) until we have an acceptable performance. Afterward, other hyper-parameters can be tuned to further increase the accuracy of the ANN, while maintaining the overall structure.

3.4.2 Activation Functions

In the early days of ANNs, the most common activation functions are identity, sigmoid, and hyperbolic tangent, shown in Fig. 3.4. Identity is used only at the output layer on regression problems so that the output of the model can be any value (positive or not). In the hidden layers, however, the use of nonlinear functions makes it possible to represent any function using ANNs, and the sigmoid and the hyperbolic tangent started being used as the activation functions because they follow the behavior of biological neurons [8].

However, in 2010, Glorot and Bengio [31] concluded that the vanishing/exploding gradient problem was caused by a poor choice of the activation function. Following this discovery, a new family of activation functions was proposed. It started with the rectified linear unit (ReLU) activation [32], as described in (3.2). The ReLU works better in deep networks because it does not saturate for positive values; besides, both

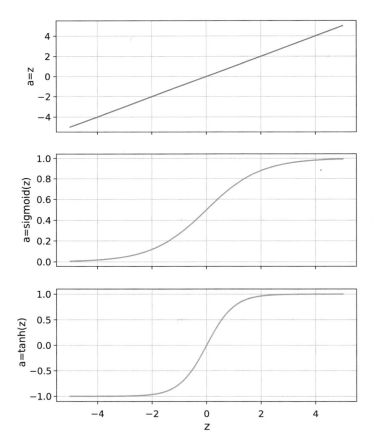

Fig. 3.4 Linear, sigmoid, and hyperbolic tangent functions

function and gradient are very easy to compute.

$$\text{ReLU}(z) = \max(0, z) \tag{3.2}$$

Still, once the sum of the weighted inputs of a neuron becomes negative, the neuron will output 0 with a gradient also of zero, which can lead to an average gradient of zero easily, once that happens those weights are no longer updated, leading to a problem known as the *dying ReLUs*. Some variants of the ReLU were proposed to address this problem. The *leaky ReLU (lReLU)* is defined according to (3.3).

$$l\text{ReLU}(z) = \max(\alpha z, z) \tag{3.3}$$

The parameter α controls the slope (leak) of the function for $z < 0$, and its value is usually close to 0. $\alpha = 0.01$ is considered a small leak, and $\alpha = 0.2$ is considered a huge leak. The *randomized leaky ReLU* acts as the *leaky ReLU,* but the parameter α is picked randomly in a given range at training time and set to the average value during testing. The *parametric leaky ReLU* turns the parameter α into a parameter that can be learned during training using the backpropagation method. The *leaky ReLU* and its variants were shown to outperform the strict ReLU function for several CNN [33].

In 2015, Clevert et al. [34] proposed a new activation function called exponential linear unit (ELU) that outperformed all the ReLU variants reducing the training time and achieving better performance on the tested problems. The ELU function is defined by (3.4). The hyper-parameter α defines the value that the function tends to when z becomes a very large negative number.

$$\text{ELU}(z) = \begin{cases} \alpha(e^z - 1) & \text{if } z < 0 \\ z & \text{if } z \geq 0 \end{cases} \tag{3.4}$$

The ELU is similar to the leaky ReLU but it is smooth for all values of z (including around $z = 0$), which helps speed up training algorithms. While it eventually saturates when $z \ll 0$, it still mitigates the occurrence of vanishing gradients problem and has a nonzero gradient in the negative spectrum, avoiding the dying units problem. The main drawback of the utilization of this activation function is that is slower to compute than the ReLU variants, which makes it slower at test time, even though it converges faster during training time. Figure 3.5 shows the ReLU, lReLU, and ELU functions.

3.4.3 How ANNs "Learn"

The learning, or training, process of an ANN corresponds to solving the optimization problem of finding the model's parameters that minimize some measure of the model's error, J. The two most common error functions are mean squared error (MSE)

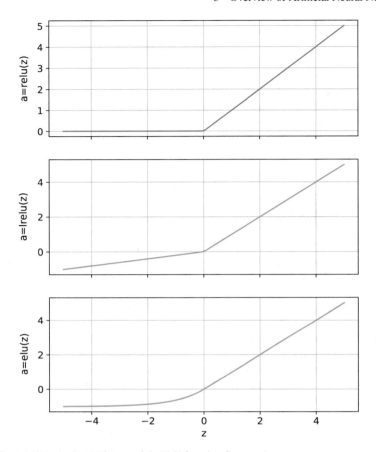

Fig. 3.5 ReLU, leaky ReLU for $\alpha = 0.2$, ELU function for $\alpha = 1$

and the mean absolute error (MAE), as defined in (3.5), where M is the number of samples, x_i is the input vector, y_i is the expected output, and w is the model's weights.

$$\text{MSE}(x, y, w) = \frac{1}{M} \sum_{i=1}^{M} (\text{ANN}(x_i, w) - y_i)^2$$

$$\text{MAE}(x, y, w) = \frac{1}{M} \sum_{i=1}^{M} |\text{ANN}(x_i, w) - y_i|$$

(3.5)

The methods used to solve this optimization problem are based on gradient descent [35], an iterative process where the cost function is travelled in the opposite direction of the gradient. In the case of ANNs' training, the gradient of the cost function with respect to the weights is computed effectively using backpropagation [5]. During training, the derivative of the error with respect to each weight is propagated backward through the network (hence the name backpropagation) and it is used to update the weights according to (3.6).

$$w^{(n+1)} = w^{(n)} - \eta \frac{\partial J}{\partial w}\left(w^{(n)}\right) \tag{3.6}$$

where J is the cost function, w is a generic weight, $w^{(n+1)}$ represents the value of that weight when it gets updated, and η is a scalar called learning step. Backpropagation exploits the chain rule and introduces an easy way to recursively computing the gradient of the cost function given the derivate of the activation functions. The backpropagation rule is represented in (3.7), where u_w is the input of the weight's w net in the feed-forward network and v_w is the input of that weight's net in the backpropagation network.

$$\frac{\partial J}{\partial w}\left(w^{(n)}\right) = u_w v_w \tag{3.7}$$

The learning rate controls the impact of the gradient on the update of the weights. A large learning rate can make the algorithm overshoot the minima of the loss function or, in the worst case, diverge from the minima. A low learning rate makes convergence too slow, as illustrated in Fig. 3.6. The learning rate is an important hyper-parameters since it controls the convergence or divergence of the optimization algorithm. However, there is a high range of values for this parameter that ensures that the optimization algorithm finds a minimum of the loss function. It is a common procedure to try larger learning rates (0.1 or 0.01) first and then starts iteratively reducing the magnitude of that value if the results improve.

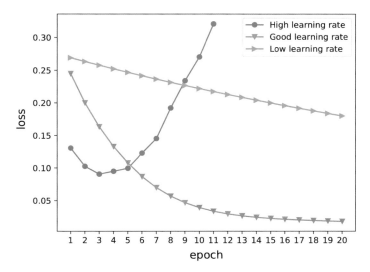

Fig. 3.6 Evolution of the error with different learning rates

As the number of samples in the data, M, becomes larger, it becomes impractical to compute the cost function over the entire dataset. In addition, by updating the weighs more often (with less computation at each time), a minimum can be found effectively. This is exactly how modern ANNs are trained, using stochastic gradient descent (SGD) [36, 37] where the weights are updated after feeding a mini-batch of examples to the network. The size of the batch impacts not only the execution time, but it also affects the properties of the minima. Large mini-batch size has been reported to cause the algorithm convergence to sharp minima, which tends to generalize badly, while small mini-batches tend to flat minimizers that tend to generalize better [38]. A large batch size is considered to be around 10% of the number of examples in the dataset and a small one is considered to be around 32, so a number between these limits should lead to acceptable results while maintaining the training time reasonable. Instead of controlling the number of updates to the weights, when training ANNs is common to measure the length of the train in the number of epochs that is the number of times, the entire training data are presented to the algorithm.

3.4.4 Optimizers

To speed up the convergence of the training process, i.e., reduce the number of iterations needed to reach a minimum of the cost function, different variants of the gradient descent were proposed and are reported to be considerably faster. These variants will be discussed in the following sections. For purposes of comparison, the update of a general weight, w, previously presented in (3.6) using the gradient descent optimization is rewritten as (3.8). The term ∂J is still computed by backpropagation. The variants of this method only change how the term Δw is computed.

$$w^{(n+1)} = w^{(n)} - \Delta^{(n+1)}w, \quad \text{where} \quad \Delta^{(n+1)}w = \eta \frac{\partial J}{\partial w}\left(w^{(n)}\right) \tag{3.8}$$

Momentum was proposed by Boris Polyark in 1964 [39], and its objective is to simulate the sense of acceleration. It takes into account the past gradients, and if the gradient keeps going in the same direction for several iterations (the same direction means the same signal), the update on the weight gets bigger as iteration progress, making it easier to escape from plateaus on the cost function. When the gradient changes directions, the algorithm reduces the step taken. The update on a weight w using momentum is given by (3.9). It introduces a new hyper-parameter β. The value of 0.9 is commonly reported to achieve good results.

$$\Delta^{(n+1)}w = \beta\Delta^{(n)}w + \eta \frac{\partial J}{\partial w}\left(w^{(n)}\right) \tag{3.9}$$

In 1983, Yurii Nesterov [40] proposed a variant of momentum optimization that almost always performs better than the vanilla version. The idea of Nesterov accelerated gradient (NAG) is to measure the gradient of the cost function not at the local position but slightly ahead in the direction of the momentum, as shown in (3.11). This tweak works since the momentum's direction usually points to the minimum of the cost function, meaning that it will accelerate the optimization process as the gradient will be evaluated closer to the optimum value.

$$\Delta^{(n+1)}w = \beta\Delta^{(n)}w + \eta\frac{\partial J}{\partial w}\left(w^{(n)} + \beta\Delta^{(n)}w\right) \tag{3.10}$$

Gradient descent starts to progress in the direction of the steepest slope, and when it reaches the end of that slope, it starts to move in the direction of the next slope. Adaptive gradient (AdaGrad) was proposed in [41], and its objective is to detect the change in the direction to the optimum value early. It achieves this by scaling down the gradient along the steepest dimension according to (3.11) where the scale factor s is computed using (3.11). As the gradient in the steepest is larger, the term s gets bigger which makes the step smaller. This is done so that the algorithm starts to move earlier in the direction of the optimum value. ε is a smoothing term to avoid division by zero, and it is usually set to 10^{-10}. An advantage of this approach is that it needs less tuning of the learning rate. AdaGrad performs well for simpler quadratic problems but can stop before reaching the minima of the cost function for complex problems like ANNs.

$$\Delta^{(n+1)}w = \frac{\eta}{\sqrt{s^{(n+1)} + \varepsilon}}\frac{\partial J}{\partial w}\left(w^{(n)}\right) \tag{3.11}$$

where

$$s^{(n+1)} = \sum_{t=1}^{n}\left[\frac{\partial J}{\partial w}\left(w^{(t)}\right)\right]^2 = s^{(n)} + \left[\frac{\partial J}{\partial w}\left(w^{(n)}\right)\right]^2 \tag{3.12}$$

Root mean square (RMS) Prop tries to solve the problem of AdaGrad stopping too early by accounting only the gradients of recent iterations instead of all the gradients since the beginning of the algorithm's execution. It uses exponential decay to compute the scaling factor s as shown in (3.13). The decay rate β is a new hyperparameter to be tuned, but setting it to 0.9 usually works well. RMSProp outperforms AdaGrad and usually performs better than momentum optimization and NAG, so it was one of the most used optimizers.

$$s^{(n+1)} = \beta s^{(n)} + (1 - \beta)\left[\frac{\partial J}{\partial w}\left(w^{(n)}\right)\right]^2 \tag{3.13}$$

Adaptive moment estimation (Adam) [42] optimization combines the ideas of momentum optimization and of RMSProp: keeps track of both an exponentially decaying average of past gradients and squared gradients (like momentum optimization and RMSProp, respectively), as shown in (3.14) and (3.15). This method introduces two new hyper-parameters to tune but usually setting β_1 to 0.9 and β_2 to 0.999 works well. As Adam is an adaptive algorithm, it needs less tuning of the learning rate η and the default value of 0.001 also works well. The existence of default values that are proven to work well makes Adam optimization is easier to use.

$$\Delta^{(n+1)}w = \frac{\eta}{\sqrt{s^{(n+1)} + \varepsilon}}m^{(n+1)} \tag{3.14}$$

where

$$
\begin{aligned}
m^{(n+1)} &= \frac{\beta_1 m^{(n)} + (1-\beta_1)\frac{\partial J}{\partial w}\left(w^{(n)}\right)}{1-\beta_1} \\
s^{(n+1)} &= \frac{\beta_2 s^{(n)} + (1-\beta_2)\left[\frac{\partial J}{\partial w}\left(w^{(n)}\right)\right]^2}{1-\beta_2}
\end{aligned}
\tag{3.15}
$$

3.4.5 Early Stop, Regularization, and Dropout

Overfitting is a major problem for large models and must be controlled. In this section, the fundamental techniques to prevent overfitting are presented. The following paragraph describes early stopping, lasso and ridge regularization, and dropout.

To evaluate how well the model will generalize, it is common to hold some of the dataset (up to 20%) from the training data and evaluate the model's predictions on this validation set. (If hyper-parameters are tuned using this validation set, then an additional test set should be reserved to verify overfitting of the hyper-parameters). Evaluating the train and validation error with the number of epochs can be used to track how well the model is doing in both datasets. The typical plot of an overfitting ANN is shown in Fig. 3.7. When the training starts, both train and validation errors decrease in a similar manner. However, after some epochs, the training error still decreases (albeit slowly) as epochs progress but validation error increases indicating that the model is overfitting to the data in the dataset.

Early stopping, as the name suggests, consists of stopping the training of the ANN as soon as there is no improvement on the validation set. To implement early stopping, one can model over the validation set at a defined interval of epochs and store snapshots of the model that has the best performance over this set. When the performance in the validation keeps on getting worse for some epochs, the training can be stopped, the model that stores used, since it is the best model in the validation set. This is a simple approach to prevent overfitting; however, it also prevents long training on complex models.

Alternatively, the lasso and ridge regularization techniques limit the value of the weights of the network, increasing sparsity in the model to prevent overfitting. Lasso

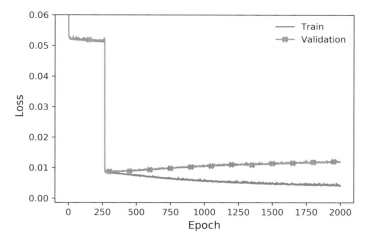

Fig. 3.7 Evolution of the training and validation errors

and ridge regularization are implemented introducing a term in the loss function used to train the network as described (3.16) and (3.17), respectively, where $J'(w)$ is the loss function used to train the network, $J(w)$ the base loss function (MAE or MSE, for example), w is the vector containing all the weights in the ANN, and λ is an hyper-parameter that controls the amount of regularization. $\|.\|_1$ represents the L_1 norm, and $\|.\|_2$ represents the L_2 norm. Due to the use of the L_1 norm and L_2 norm, these regularization techniques are often L_1 and L_2 regularization, respectively.

$$J'(w|x, y) = J(w|x, y) + \lambda \sum |w_i| = J(w|x, y) + \lambda \|W\|_1 \qquad (3.16)$$

$$J'(w|x, y) = J(w|x, y) + \lambda \sum w_i^2 = J(w|x, y) + \lambda \|W\|_2^2 \qquad (3.17)$$

Yet, another approach to prevent overfitting is dropout [43]. Dropout consists of dropping out neurons in a neural network, by temporarily removing it from the network as shown in Fig. 3.8. When dropout is applied to a network, the original is sampled and the result is a simpler network as represented in Fig. 3.8b. This sample of the network contains only the units that survived dropout, and so an ANN with n units corresponds to 2^n simpler networks (because units can be active or dropped out). At training time, for each training step the network is sampled and trained. In this sense, applying dropout to a network is equivalent to train 2^n simpler networks with different architectures, where each of these simpler networks gets trained very rarely, if at all. The combination of 2^n networks with different architectures into a single network that outputs an (approximation of the) average of their outputs results in a significantly lower generalization error on a wide variety of problems [43].

The probability of keeping each node, p, is a new hyper-parameter that controls how much dropout is applied. A 50% probability of a unit to be active is commonly

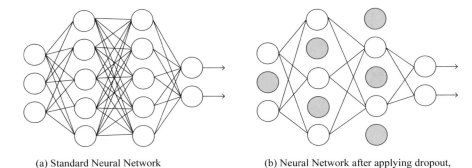

(a) Standard Neural Network   (b) Neural Network after applying dropout,
                                            grayed units represent dropped out units.

Fig. 3.8 Standard ANN with and without dropout

used. Before using the network to make predictions, the outgoing weights each unit
are multiplied by the probability, p, used during training.

3.5 Summary

This chapter peruses the main branches of machine learning and how the different
techniques have found their way into analog EDA. ANNs and deep learning were
shown to be usable for analog IC modeling and synthesis and are further explored in
this work

ANNs are very flexible models that can be easily tailored for the target problem
and have the ability to mix both regression and classification tasks in one single
model. Additionally, they are appropriate for a high throughput of data. While their
computation times may vary with the amount of data being processed and become
quite long, this will not affect posterior predictions, since training is the most time-
consuming task. The cost of ANNs flexibility is that there are a plethora of hyper-
parameters to be tuned, whose best values depend on the problem and hand and can
only be found by trial and error. In the last section of this chapter, the most relevant
hyper-parameters were presented as well as some strategies to select and train the
best model.

References

1. P. Langley, The changing science of machine learning. Mach. Learn. **82**, 275–279 (2011)
2. T. Bayes, An essay towards solving a problem in the doctrine of chances. Phil. Trans. **53**,
   370–418 (1763), https://doi.org/10.1098/rstl.1763.0053
3. P. Diaconis, The Markov chain Monte Carlo revolution. Bull. Am. Math. Soc. **46**, 179–205
   (2009)

4. F. Rosenblatt, The perceptron: a probabilistic model for information storage and organization in the brain. Psychol. Rev. **65**(6), 386–408 (1958)
5. D. Rumelhart, G. Hinton, R. William, Learning representations by back-propagating errors Nature **323**(9), 533–536 (1986)
6. M.I. Jordan, T.M. Mitchell, Machine learning: trends, perspectives and prospects. Science **349**, 255–260 (2015)
7. J. VanderPlas, Machine learning, in *Python data science handbook essential tools for working with data* (O'Reilly Media, 2016), p. 541
8. A. Géron, *Hands-on Machine Learning with Scikit-Learn & TensorFlow* (O'Reilly, 2017)
9. AlphaGo Zero: Learning from Scratch, 2017. https://deepmind.com/blog/alphago-zero-learningscratch/. Accessed 4 Oct 2019
10. M. Banko, E. Brill, Scaling to very very large corpora for natural language disambiguation, in *Proceedings of the 39th Annual Meeting on Association for Computational Linguistics—ACL '01* (2001), pp. 26–33
11. P. Domingos, *The Master Algorithm: How the Quest for the Ultimate Learning Machine Will Remake Our World* (Basic Books, New York, 2015)
12. S. Karamizadeh, S.M. Abdullah, M. Halimi, J. Shayan, M.J. Rajabi, Advantage and drawback of support vector machine functionality, in *International Conference on Computer, Communication, and Control Technology* (2014)
13. J.A. Vrugt, B.A. Robinson, Improved evolutionary optimization from genetically adaptive multimethod search. PNAS **104**, 708–711 (2006)
14. D.B. Fogel, *The Advantages of Evolutionary Computation* (Natural Selection Inc, 1997)
15. R. Martins, N. Lourenço, F. Passos, R. Póvoa, A. Canelas, E. Roca, R. Castro-López, J. Sieiro, F.V. Fernández, N. Horta, Two-step RF IC block synthesis with pre-optimized inductors and full layout generation in-the-loop. IEEE Trans. Comput. Aided Des. Integr. Circuits Syst. (TCAD) **38**(6), 989–1002 (2019)
16. N. Lourenço, R. Martins, N. Horta, *Automatic Analog IC Sizing and Optimization Constrained with PVT Corners and Layout Effects* (Springer, 2017)
17. R. Martins, N. Lourenco, A. Canelas, N. Horta, Electromigration-aware and IR-drop avoidance routing in analog multiport terminal structures, in *Design, automation & test in Europe conference (DATE)* (Dresden, Germany, 2014), pp. 1–6
18. R. Martins, N. Lourenço, N. Horta, J. Yin, P. Mak, R. Martins, Many-objective sizing optimization of a class-C/D VCO for ultralow-power IoT and ultralow- phase-noise cellular applications. IEEE Trans. Very Large Scale Integr. (VLSI) Syst. **27**(1), 69–82 (2019)
19. F. De Bernardinis, M. I. Jordan, A. SangiovanniVincentelli, Support vector machines for analog circuit performance representation, in *Proceedings 2003. Design Automation Conference* (IEEE Cat. No. 03CH37451) (Anaheim, CA, 2003)
20. N. Lourenço, R. Martins, M. Barros, N. Horta, Chapter in analog/RF and mixed-signal circuit systematic design, in *Analog Circuit Design based on Robust POFs using an Enhanced MOEA with SVM Models,* ed. by M. Fakhfakh, E. Tielo-Cuautle, R. Castro-Lopez (Springer, 2013), pp 149–167
21. J. Tao et al., Large-scale circuit performance modeling by bayesian model fusion, in *Machine Learning in VLSI Computer-Aided Design*, ed. by I. Elfadel, D. Boning, X. Li (Springer, Cham, 2019)
22. W. Lyu, F. Yang, C. Yan, D. Zhou, X. Zeng, Multi-objective Bayesian optimization for analog/RF circuit synthesis, in *2018 55th ACM/ESDA/IEEE Design Automation Conference (DAC)* (San Francisco, CA, 2018)
23. J. Schmidhuber, Deep learning in neural networks: an overview. Neural Netw. **61**, 85–117 (2015)
24. K. Hornik, M. Stinchcombe, H. White, Multilayer feedforward networks are universal approximators. Neural Netw. **2**(5), 359–366 (1989)
25. Y. LeCun, Y. Bengio, G. Hinton, Deep learning. Nature **521**(7553), 436–444 (2015)
26. K. He, X. Zhang, S. Ren, J. Sun, Deep residual learning for image recognition, in 2016 *IEEE Conference on Computer Vision and Pattern Recognition (CVPR)* (Las Vegas, NV, 2016), pp. 770–778

27. R. Eldan, O. Shamir, The power of depth for feedforward neural networks, in *Conference on Learning Theory* (2016), pp 907–940
28. G.B. Huang, Learning capability and storage capacity of two-hidden-layer feedforward networks. IEEE Trans. Neural Netw. **14**(2), 274–281 (2003)
29. K. Jinchuan, L. Xinzhe, Empirical analysis of optimal hidden neurons in neural network modeling for stock prediction, in *Proceedings-2008 Pacific-Asia Workshop on Computational Intelligence and Industrial Application, PACIIA 2008*, vol. 2 (2008), pp. 828–832
30. S. Trenn, Multilayer perceptrons: approximation order and necessary number of hidden units. IEEE Trans. Neural Netw. **19**(5), 836–844 (2008)
31. X. Glorot, Y. Bengio, Understanding the difficulty of training deep feedforward neural networks, in *Proceedings of the Thirteenth International Conference on Artificial Intelligence and Statistics*, vol. 9 (2010), pp. 249–256
32. V. Nair, G.E. Hinton, Rectified linear units improve restricted boltzmann machines, in *Proceedings of the 27th International Conference on International Conference on Machine Learning* (2010)
33. B. Xu, N. Wang, T. Chen, M. Li, Empirical evaluation of rectified activations in convolutional network, in *ICML Deep Learning Workshop* (Lille, France, 2015)
34. D.A. Clevert, T. Unterthiner, S. Hochreiter, Fast and accurate deep network learning by Exponential Linear Units (ELUs), in *International Conference on Learning Representations* (2015), pp. 1–14
35. A. Cauchy, Méthode générale pour la résolution des systemes d'équations simultanées. Comp. Rend. Sci. Paris **25**, 536–538 (1847)
36. H. Robbins, S. Monro, A stochastic approximation method the annals of mathematical statistics. An. Math. Stat. **22**(3), 400–407 (1951)
37. J. Kiefer, J. Wolfowitz, Stochastic estimation of the maximum of a regression function. Ann. Math. Stat. **23**(3), 462–466 (1952)
38. N.S. Keskar, D. Mudigere, J. Nocedal, M. Smelyanskiy, P.T.P. Tang, On large-batch training for deep learning: generalization gap and sharp minima, in *5th International Conference on Learning Representations, ICLR 2017—Conference Track Proceedings* (2017)
39. B. Polyak, Some methods of speeding up the convergence of iteration methods. USSR Comput. Math. Math. Phys. **4**, 1–17 (1964)
40. Y. Nesterov, A method for solving the convex programming problem with convergence rate $O(1/k^2)$. Dokl. Akad. Nauk SSSR **269**, 543–547 (1983)
41. J. Duchi, E. Hazan, Y. Singer, Adaptive subgradient methods for online learning and stochastic optimization. J. Mach. Learn. Res. **12**, 2121–2159 (2011)
42. D.P. Kingma, J.B, Adam: a method for stochastic optimization. CoRR, abs/1412.6980 (2014)
43. N. Srivastava, G. Hinton, A. Krizhevsky, I. Sutskever, R. Salakhutdinov, dropout: a simple way to prevent neural networks from overfitting. J. Mach. Learn. Res. **15**, 1929–1958 (2014)

Chapter 4
Using ANNs to Size Analog Integrated Circuits

4.1 Design Flow

For automatic sizing, simulation-based optimization approaches are the most prevalent methods in both industrial [1, 2] and academic [3, 4] environments, and the consideration of variability [5] models from electromagnetic simulation [6] or layout effects [7] is nowadays a reality. Still, these processes are extremely time-consuming, which has led to the use of machine learning, and, in particular, ANN models, to address analog IC sizing automation. These models were usually used to estimate the performance parameters of complete circuit topologies [8, 9] or migrate basic circuit structures for different integration technologies [10]. In a different direction, in this chapter, ANNs [11] are used for automatic analog IC sizing [12] by reusing data available from previous design runs. The following steps summarize the stages of the proposed design flow.

1. Determine the prediction target.
2. Collect data for learning.
3. Create learning model.
4. Train learning model.
5. Confirm the accuracy of the prediction.
6. Sample the obtained results.
7. Simulate the sampled results on the circuit simulator.

The first step in the creation of an ANN model is to determine the prediction target. In this case, we want the network to learn design patterns from the studied circuits, using circuit's performances (DC gain, current consumption (IDD), gain bandwidth (GBW), and phase margin (PM)) as input features and devices' sizes (such as such as widths and lengths of resistors and capacitors) as target outputs. The next step involves gathering data so that the model can learn patterns from the input–output mapping. Ideally, this data should be reused from previous designs. Data should be split into three different sets: the training set, the validation set, and

© The Author(s), under exclusive license to Springer Nature Switzerland AG 2020
J. P. S. Rosa et al., *Using Artificial Neural Networks for Analog Integrated Circuit Design Automation*, SpringerBriefs in Applied Sciences and Technology, https://doi.org/10.1007/978-3-030-35743-6_4

the test set. After determining the prediction target and assembling the data, hyper-parameters of the model should be selected. These are the number of layers, number of nodes per layer, the activation functions, and the loss function, to name a few. After selecting the most appropriate hyper-parameters, the model is ready to be trained. At this stage, the model will attempt to iteratively find its optimal network weights. After training the model and achieving a sufficiently low error on the training phase and on the validation set, the model is ready to be used for predicting the output of real-world input data. This is where we will evaluate the accuracy of the predicted results and obtain devices' sizes from the test set. The next step involves sampling of the results from the model. This step is essential to circumvent possibly biased solutions predicted by the ANN. In the final step, we verify the usefulness of the obtained results using the circuit simulator.

4.2   Problem and Dataset Definition

Let $V_{(i)} \in \mathbb{R}^N$ be the N-dimensional array of design variables that define the sizing of the circuit, where the index i inside the chevron identifies solution point i in the dataset, and $S_{(i)} \in \mathbb{R}^D$ be the D-dimensional array of the corresponding circuit performance figures. The ANNs presented in this work are in general trained to predict the most likely sizing given the target specifications according to (4.1).

$$V^* \sim \arg \max(P(V|S)) \tag{4.1}$$

Hence, to train the ANN, the training data are comprised of a set T of M data pairs $\{V, S\}_{(i)}$. Since we want the model to learn how to design circuits properly, these pairs must correspond to useful designs, e.g., optimal or quasi-optimal design solutions for a given sizing problem.

Up until now, the definition in (4.1), like in other state-of-the-art approaches, allows to train a model suitable for analog IC sizing. However, if nothing is done, the model will only predict reasonable designs if imputed with target specifications vales that can be met all at once by the circuit. However, that is not usually the use case since circuits' target specifications are, more often than not, defined as inequalities instead of equalities. Therefore, an ANN trained to map $S \rightarrow V$ may have difficulties predicting to specification values that are actually worse than the ones provided in training.

The point is, if the sizing $V_{(i)}$ corresponds to a circuit whose performance is $S_{(i)}$, then it is also a valid design for any specifications whose performance targets, $S'_{(i)}$, are worse than $S_{(i)}$. Having this in mind, an augmented dataset T' can be obtained from T as the union of T and K modified copies of T, as indicated in (4.2), where the for each sample i, the $S_{(i)}$ is replaced by $S'_{(i)}$ according to (4.3).

$$T' = \{T \cup T^{C1} \cup T^{C2} \cup T^{CK}\} \tag{4.2}$$

$$S'_{<i>} = S_{<i>} + \left(\frac{\gamma}{M} \sum_{j=1}^{M} S_{<j>} \right) \Delta \Gamma \tag{4.3}$$

where $\gamma \in]0, 1[$ is a factor used to scale the average performances, Δ is a diagonal matrix of random numbers between $[0, 1]$, and $\Gamma \in \{-1, 1\}$ is the target diagonal matrix that defines the scope of usefulness of a circuit performance. Its diagonal components take the value -1 for performance figures in which a smaller target value for the specification is also fulfilled by the true performance of the design, e.g., DC gain, and it takes the value 1 is for the diagonal components corresponding to performance figures that meet specification targets that are larger than the true performance of the circuit, e.g., power consumption.

The data are scrabbled and split into three separate sets:

- **Training Set**: Set of examples used for learning. In the case of an ANN, the training set is used to find the optimal network weights with backpropagation.
- **Validation Set**: Set of examples used to tune model hyper-parameters. Performance metrics such as mean squared error (MSE) or mean absolute error (MAE) can be applied to the classification of this set of data to evaluate the performance of the model.
- **Test Set**: Set of examples used to evaluate the performance of a fully trained classifier on completely unseen real-world data, after the model has been chosen and fine-tuned.

This segmentation is one the first measures a designer can take to avoid overfitting. If the model would be tested on the same data it was trained with, the obtained classification would yield no error because the test data would have already been presented to the model. Therefore, no meaningful conclusions regarding the effectiveness of the model could be made. To avoid this, part of the available data are held out and used as validation data to test the model on unseen examples so that the designer can evaluate performance metrics and fine-tune model hyper-parameters. After the model has been finalized, a test set is fed to the fully trained classifier to assess its performance. Validation and test sets are separate because the error rate estimate of the final model on validation data is usually biased (smaller than the true error rate) since the validation set is used to select the final model [11]. The percentage assigned to each set is variable, but it is common to use 60, 20, and 20% of total data for the training, validation, and test sets, respectively.

If the data are scarce, simply partitioning the data into three distinct sets might affect the performance of the model in the training phase. In this case, other methods might be more appropriate for partitioning the data. These alternative procedures are known as cross-validation methods. Although they come at a higher computational cost, cross-validation methods are ideal for small datasets. A test set should still be held out for final evaluation, but the need for a validation set vanishes. A particular method of cross-validation called *k-fold cross-validation* is illustrated in Fig. 4.1 and

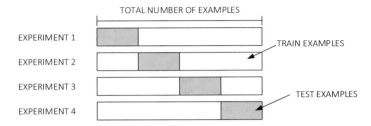

Fig. 4.1 K-fold cross-validation

consists of splitting the training set into k smaller sets. The advantage of k-fold cross-validation is that all the data selected for the training set are used for both training and validation. For each of the k-folds, the following procedure is applied:

1   The model is trained using $k-1$ of the folds of the training data.
2   The remainder of the data is used to validate the model and compute performance metrics.
3   The true error is estimated as the average error rate on validation data of all k-folds.

4.3   Regression–Only Model

The ANNs models discussed in this section consider fully connected layers without weight sharing [11]. Given the number of features that are used in the model and the size of the datasets that will be considered in this application, the model is not very deep, containing only a few of hidden layers.

The input features measure different attributes from the circuits, so it is recommended to standardize the input matrix. A scaler is applied after we expand the feature space by generating polynomial features, which is quite useful since we only consider four circuit performances as input features. Adding extra columns of features helps giving more flexibility to the model.

More specifically, the ANN is trained with an input X that is the feature mapping Φ of S normalized. Each input data sample $X_{\langle i \rangle}$ is given by (4.4), where $\Phi(S_{\langle i \rangle})$ is a second-order polynomial mapping of the original specifications, e.g., for $S_{\langle i \rangle} = [a, b, c]$, $\Phi(S_{\langle i \rangle}) = [a, b, c, a^2, ab, ac, b^2, bc, c^2]$; μ_Φ is the mean of the components of Φ, and σ_ϕ is the standard deviation of the components of Φ. Polynomial features degree is a hyper-parameter of the model. For the examples presented in this chapter, it was chosen to be of second order and proved to be effective. Higher-order polynomial features were tested, but no considerable improvements observed.

$$X_{\langle i \rangle} = \frac{\Phi(S_{\langle i \rangle}) - \mu_\Phi}{\sigma_\Phi} \tag{4.4}$$

The output of the network, Y, is defined from V by (4.5).

$$Y_{\langle i \rangle} = \frac{V_{\langle i \rangle} - \min(V)}{\max(V) - \min(V)} \tag{4.5}$$

The base of the network architecture design is the input layer and an output layer whose size is determined by the dataset. The number of hidden layers is adjusted to evaluate model complexity. The best structure depends of the dataset, but a systematic method is proposed later in this section to specify such models. To train and evaluate the model, the datasets are split in training (80–90%) and validation sets (20–10%). An additional small test set, whose specifications are drawn independently, is used to verify the real-world application of the model. A base structure for the model is shown in Fig. 4.2.

The number of features and design variables of the circuit define the number of input and output nodes. Only the number of hidden layers and its nodes need tuning to obtain an architecture yielding lower training and validation error. The methodology that was used to select the hyper-parameters for the ANN regression-only model architectures was as follows:

1. The number of input nodes is obtained from the second-order polynomial feature extension of the circuits' performance figures. Models with up to five hidden layers were experimented. The number of selected hidden layers was 3. As a rule of thumb, the number of nodes should increase in the first hidden layer to create a rich encoding (120 nodes) and then decreases toward the output layer

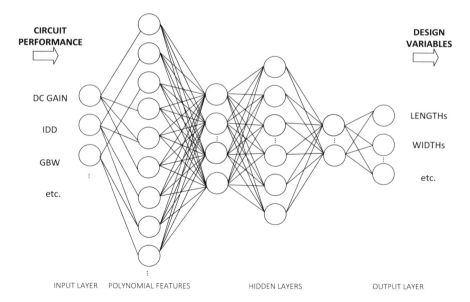

Fig. 4.2 Base structure of the regression-only model

   to decode the predicted circuit sizing (240 and 60 nodes in the second and third
   hidden layers, respectively). The number of output nodes depends on the number
   of devices' sizes from the topology being studied.

2. In initial testing, sigmoid was used as the activation function of all nodes from
   all layers, except the output layer, but ReLU [13] ended up being the preferred
   choice. Mainly, the gradient using ReLU is better propagated while minimizing
   the cost function of the network. Output layer nodes do not use activation functions because we wish to find the true values predicted by the network instead of
   approximating them to either end of an activation function.

3. Initially, the stochastic gradient descent optimizer was used, but later dropped
   in favor of Adam, which has better documented results. Parameters chosen for
   Adam such as learning rate are the ones recommended in the original publication
   [14].

4. The models were first designed to have good performance (low error) in the
   training data, even if overfitting was observed. This method allowed to determine
   the minimum complexity that can model the training data. Overfitting was then
   addressed using L_2 weight regularization, as shown in Fig. 4.3.

5. Initial random weights of the network layers were initialized by means of a
   normal distribution.

6. Model performance was measured through a MAE loss function, while the
   training of the model was performed using a MSE loss function.

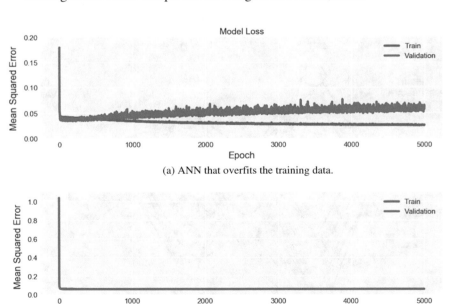

(a) ANN that overfits the training data.

(b) Same ANN trained with L2 regularization.

Fig. 4.3 Evolution of train and validation errors during training

Table 4.1 Hyper-parameters for the regression-only model

Hyper-parameter	Name/value
Input layer	1 layer (15 nodes)
Hidden layers	3 layers (120, 240, 60 nodes, respectively)
Output layer	1 layer (12 nodes for VCOTA—amplifier using voltage combiners for gain enhancement—topology or 15 nodes for two-stage Miller topology)
Activation function	ReLU
Optimizer	Adam (l earning rate = 0.001)
Kernel regularizer—L_2	lambda = 1.3E−05 (VCOTA) lambda = 2.0E−05 (two-stage Miller)
Kernel initializer	Normal distribution
Loss function	MSE
Accuracy metrics	MAE
Number of epochs used for training	500–5000
Batch size used for training	256–512

7. A high number of epochs (5000) were chosen for the training of early networks to ensure that the training reached the lowest possible error. One epoch occurs when all training examples execute a forward pass and a backward pass through the network. It is often a good practice to train the network for a high number of epochs, save the network weights from the training, and perform future testing using those weights with a lower number of epochs (e.g., 500).
8. A variable number was chosen for batch size, between 32 and 512, depending on the number of epochs. Batch size is the number of training examples in one forward pass/ backward pass.
9. Finally, *gridsearch* is done over some hyper-parameters (number of layer, number of nodes per layer, non-ideality, and regularization factor) to fine-tune the model [11].

The hyper-parameters chosen for this architecture are summarized in Table 4.1.

4.3.1 Training

The loss function, L_1, of the model that is optimized during training is the MSE of the predicted outputs Y' with respect to the true Y plus the L_2 norm of the model's weights, W, times the regularization factor λ, according to (4.6).

$$L_1 = \frac{1}{M} \sum_{j=1}^{M} \left(\left(Y'_{(j)} - Y_{(j)} \right)^T \left(Y'_{(j)} - Y_{(j)} \right) \right) + \lambda \| W \|^2 \qquad (4.6)$$

Training of the models is done using the Adam optimizer [14], a variant of stochastic steepest descent with both adaptive learning rate and momentum that provides good performance. Moreover, it is quite robust with respect to its hyper-parameters, requiring little or no tuning. It is well suited for problems that are large in terms of data and appropriate for problems with noisy or sparse gradients. Adam's configuration parameters are as follows:

- **alpha**: Also referred to as the learning rate or step size, i.e., the proportion in which weights are updated. Larger values result in faster initial learning before the rate is updated. Smaller values result in slower learning during training.
- **beta1**: The exponential decay rate for the first moment estimates.
- **beta2**: The exponential decay rate for the second moment estimates. This value should be set close to 1.0 on problems with a sparse gradient.
- **epsilon**: Parameter that should be as close to zero as possible to prevent any division by zero in the implementation.
- **decay**: Learning rate decay over each update.

Authors of the paper proposing Adam suggest using alpha $= 0.001$, beta1 $= 0.9$, beta2 $= 0.999$, epsilon $= 10^{-8}$, and decay $= 0.0$ as parameters.

When training a supervised learning model, the main goal is to achieve an optimal generalization of unseen data. When a model achieves low error on training data but performs much worse on test data, we say that the model has *overfit*. This means that the model has caught very specific features of the training data, but did not really learn the general patterns. The best way to evaluate this behavior is through error analysis.

Test error is also commonly referred to as generalization error because it reflects the error of the model when generalized to previously unseen data. When we have simple models and abundant data, we expect the generalization error to resemble the training error. When we work with more complex models and fewer examples, we expect the training error to go down but the generalization gap to grow. Some factors that may affect generalization error are: the number of hyper-parameters, the values taken by them, and the number of training examples.

Model complexity is governed by many factors, and its definition is not straightforward. For example, a model with more parameters might be considered more complex. A model whose parameters can take a wider range of values might be more complex. Often with neural networks, it is common to think of a model that takes more training steps as more complex, and one subject to *early stopping* as less complex. Complexity is no exact science, though, but it can be loosely defined by models that can readily explain *arbitrary* facts, whereas models that only have a limited expressive power but still manage to explain the data well are probably closer to the truth.

L_2 regularization was used in the models proposed in this chapter, as seen in (4.6). This is a technique that helps overcoming overfitting and is a regularization term that is added to the loss function in order to prevent coefficients to fit perfectly and undergeneralize.

The use of L_2 regularization proved to be effective in the case study. Initial testing cases, where only input and output layers were considered, did not achieve overfitting in the validation set because the complexity of the model was not high enough to produce that effect. When hidden layers were later added with a variable number of nodes, model complexity started to continually increase until overfitting became apparent, a behavior that can be observed in Fig. 4.3a. By adding the L_2 regularization term, overfitting was completely negated, which can be observed in Fig. 4.3b.

4.3.2 Using the ANN for Circuit Sizing

Obtaining a sized circuit from the ANN is done using (4.3) P times, with Γ replaced by $-\Gamma$; i.e., we ask the model to predict a set of P sizing solutions for target circuit performances that are better than the desired specifications. In case not all performance figures used to train the model are specified, then the corresponding component in the diagonal random matrix Δ should be a random value in the range of $[-1, 1]$. For instance, if the target specifications are gain bandwidth product (GBW) over 30 MHz and current consumption (IDD) under 300 μA, and the model was trained with DC gain, GBW, and IDD. A set of circuit performances given to the ANN could be, e.g., {(50 dB, 35 MHz, 290 μA), (75 dB, 30 MHz, 285 μA), (60 dB, 37 MHz, 250 μA), … (90 dB, 39 MHz, 210 μA)}.

The reasoning behind this sampling is that even if the ANN has properly learned the designs patterns present in the performances of the sizing solutions in the training data, when the performance trade-off implied by the target specifications being requested is not from the same distribution than the training data, the prediction of the ANN can be strongly and badly biased. While using the augmented dataset described in Sect. 4.2 alleviates this bias, it is still better to sample the ANN this way. Another reason for this sampling is that a given set of predictions might not be accurate enough for some devices' sizes. Specific values of design variables from the sizing output might be very far from the ones desired, and sampling is a good way to circumvent this problem.

The selection of solutions from the P predictions of the ANN is done by simulating the predicted circuit sizing, and either: use a single value metric such as some figure-of-merit (FoM) to select the most suitable solution or using some sort of Pareto dominance to present a set of solutions exploring the trade-off between specifications.

4.4 Classification and Regression Model

The ANN architecture considered in this section is similar to the one used for the regression-only model, but now there is an increased number of output nodes, as shown in Fig. 4.4. The input features are now not only restricted to one class of circuits, but to three. The features still correspond to the same four performance

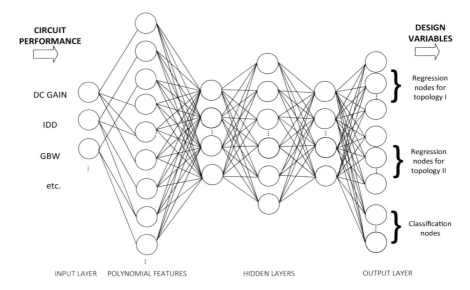

Fig. 4.4 Base structure of the classification and regression model

measures used in the regression-only model. The output layer is now not only comprised of a series of nodes that represent the circuit's sizes, but also an additional node for each class of circuits present in the dataset. The loss function used in the training of the networks will also be different, now taking into account both errors from the regression and the classification tasks. The weights assigned to each error measures are malleable, but weights of 70% and 30%, respectively, were used as a starting point.

For this case study, three classes of circuits were considered. The output nodes responsible for classification assign a probability to the predicted class. Data preparation for this model involved the same steps of normalization and data augmentation through polynomial features as the ones from the regression-only model. The guidelines to select the hyper-parameters for the ANN classification and regression model architecture were as follows:

1. The number of input nodes and the number of hidden layers are the same as the regression-only model architecture. The number of nodes in the output layer increases in relation to the previous model, which are now 30. This reflects the fact that the network is now processing different circuit performances and target circuit measures: 12 nodes for the VCOTA topology and 15 nodes for the two-stage Miller amplifier topology, and 3 additional nodes that encode the circuit class.
2. The activation function used in all nodes (except in the output layer' nodes) is ReLU.
3. Adam was the chosen optimizer, with learning rate = 0.001.
4. Overfitting was addressed using L_2 weight regularization, as shown in Fig. 4.5,

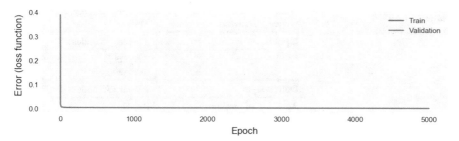

Fig. 4.5 Train and validation error during training with L_2 regularization

after the model showed to have a good performance.

5. Initial random weights of the network layers were initialized by means of a normal distribution.

6. Model performance was measured through a custom loss function (see Eq. 4.8) that takes into account the error measurements from the classification nodes and from the regression nodes. Different percentages are assigned to each type of error, 30% and 70%, respectively. Individual metrics were also used to prove the effectiveness of each task in the network. Regression error is calculated though a MSE function, while classification error is calculated through a sparse softmax cross-entropy (SSCE) function.

7. 5000 was the number of epochs chosen for initial testing. After having trained the first model, subsequent ANNs were trained with fewer epochs (500), using network weights from the ANN trained for 5000 epochs.

8. A variable number was chosen for batch size, between 256 and 512, depending on the number of epochs.

9. Finally, *gridsearch* is once again done over the hyper-parameters (number of layer, number of nodes per layer, non-ideality, and regularization factor) to fine-tune the model.

The hyper-parameters chosen for this architecture are summarized in Table 4.2.

4.4.1   Training

The loss function, L_2, of the model that is optimized during training is a weighted sum of two distinct losses—one from the regression task and the other from the classification task. Since this model's input features are not restricted to only one class of circuit performances, the regression loss will itself be a sum of the training errors from each circuit included in the dataset. Each individual regression loss is determined using MSE, like the previous model, while the classification error is measured through a SSCE function. This function measures the probability error in discrete classification tasks in which the classes are mutually exclusive (each entry is in exactly one class).

Table 4.2 Model
hyper-parameters

Hyper-parameter	Value
Input layer	1 layer (15 nodes)
Hidden layers	3 layers (120, 240, 60 nodes, respectively)
Output layer	1 layer (36 nodes)
Activation function	ReLU
Optimizer	Adam (learning rate = 0.001)
Kernel regularizer	L_2 (lambda = 2.0E−05)
Kernel initializer	Normal distribution
Loss function	Custom (see expression (30))
Accuracy metrics	MAE, SSCE
Number of epochs used for training	500–5000
Batch size used for training	256–512

The loss function, L_{class}, that is optimized for the classification task is obtained by computing the negative logarithm of the probability of the true class, i.e., the class with highest probability as predicted by the ANN:

$$L_{\text{class}} = -\log(p(Y_{\text{class}})) \tag{4.7}$$

The loss function, L_{reg}, that is optimized for the regression task is the MSE of predicted outputs Y' with respect to the true Y plus the L_2 norm of the model's weights, W, times the regularization factor λ:

$$L_{\text{reg}} = \frac{1}{M} \sum_{j=1}^{M} \left(\left(Y'_{(j)} - Y_{(j)} \right)^T \left(Y'_{(j)} - Y_{(j)} \right) \right) + \lambda \| W \|^2 \tag{4.8}$$

The total loss function, L_2, is the weighted sum between the two previous loss functions. Since there are two classes of circuits (excluding the third one, which is ignored in this function), there will be a distinct loss function value from each regression applied to each class. The MSE from each class is then multiplied by the true class predicted by the network in each step. This means that the MSE for the other class that was not predicted will be neglected and become zero; i.e., if for a given step, VCOTA is the predicted topology, TrueClass$_1$ will be greater than zero, while TrueClass$_2$ will be equal to zero. The formulation of L_2 is as follows:

$$L_2 = 0.30 \times L_{\text{class}} + (0.70 \times (L_{\text{reg1}} \times Y_{\text{class1}} + L_{\text{reg2}} \times Y_{\text{class2}})) \tag{4.9}$$

The training of the models is again done using the Adam optimizer [14]. Other error metrics such as MAE and SSCE are also considered when validating the results. The results below are obtained for a model with one input and one output layer,

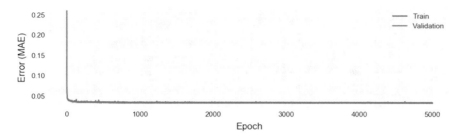

Fig. 4.6   Model regression error

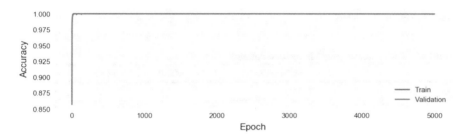

Fig. 4.7   Model classification error

and three hidden layers with 120, 240, 60 nodes each, for the demonstration of L_2 regularization effectiveness.

Similar to the previous architecture, model loss did not show overfitting after L_2 regularization was included, as shown in Fig. 4.5. In this architecture, regression is performed using the same functions as the previous architecture: MSE for the training of the network and MAE for error measurement. Thus, the error obtained is similar to the one obtained in the regression-only model, as shown in Fig. 4.6. Figure 4.7 shows that classes from the all the design points are correctly predicted on the validation set.

4.5   Test Case–Regression–Only Model

For proof of concept, the amplifier using voltage combiners for gain enhancement (VCOTA) from [15] was used, and, for a second example, a two-stage Miller amplifier was considered. The circuits' schematics are shown in Fig. 4.8.

The goal of this architecture is to learn design patterns from the studied circuits. By mapping from the device's sizes to the circuit's performances, using regression, the networks should be able to predict new devices' sizes by using as input features the desired circuit performances. Datasets were obtained from a series of previously

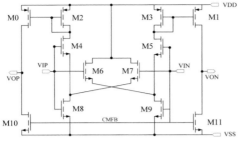
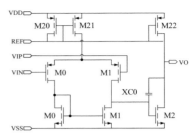

(a) Single stage amplifier with gain enhancement using voltage combiners

(b) Two Stage Miller amplifier

Fig. 4.8 Circuit schematics

Table 4.3 Performance ranges in the two datasets

		DC gain (dB)	GBW (MHz)	IDD (μA)	PM (°)
VCOTA	Max	56.8	78	395	80
	Min	44.7	34	221	60
Two-stage Miller	Max	97.2	102	0.8	90
	Min	59.8	1.5	0.3	55

done studies on this circuit for the UMC 130 nm technology design process and contain only optimized circuit sizing solutions.

For proof of concept, the first study was performed on the VCOTA topology. Dataset for this example, i.e., Dataset-1, before any augmentation has 16,600 different design points. The second study was performed on the two-stage Miller topology. For this example, the dataset before any augmentation, i.e., Dataset-2, has 5162 different design points. The circuit performances that were considered to train the ANN in both cases were DC gain, IDD, GBW, and PM, and the ranges of values found in the dataset are shown in Table 4.3.

4.5.1   Single-Stage Amplifier with Voltage Combiners

Three ANNs were trained for the VCOTA topology, i.e., ANN-1 to ANN-3. All ANNs were implemented in Python with Keras and TensorFlow [16] as backend. The code was run, on an Intel® Core™ i7 Quad CPU 2.6 GHz with 8 GB of RAM. The structure considered has 15 input variables (obtained from the second-order polynomial feature extension of the 4 performance figures from Table 4.3), 3 hidden layers with 120, 240, 60 nodes each, and the output layer has 12 nodes.

ANN-1 was trained on the original dataset, for 5000 epochs with batches of 512 samples. Its training took less than 15 min. ANN-2 was trained on the dataset

Table 4.4 Performance of the ANNs trained for the VCOTA

		MSE train	MSE Val.	MAE train	MAE Val.
VCOTA	ANN-1	0.0159	0.0157	0.0775	0.0776
	ANN-2	0.0124	0.0123	0.0755	0.0750
	ANN-3	0.0124	0.0124	0.0754	0.0753

Table 4.5 MAE of device's sizes for the VCOTA

Transistor width	MAE			Transistor length	MAE		
	ANN-1	ANN-2	ANN-3		ANN-1	ANN-2	ANN-3
W0	1.06E−05	1.84E−06	1.83E−06	L0	2.11E−08	5.08E−08	5.07E−08
W1	8.91E−07	2.22E−06	2.17E−06	L1	5.32E−08	9.73E−08	9.98E−08
W4	5.24E−06	6.05E−06	6.04E−06	L4	1.21E−07	1.37E−08	1.36E−08
W6	1.48E−05	1.62E−06	1.62E−06	L6	1.39E−07	1.56E−08	1.56E−08
W8	2.75E−09	2.70E−09	2.65E−09	L8	2.98E−08	3.41E−08	3.35E−08
W10	4.33E−06	4.59E−06	4.51E−06	L10	6.61E−08	6.99E−08	7.05E−08

W0 and L0 are the width and length of the transistors M0 and M1
W2 and L2 are the width and length of the transistors M2 and M3
W4 and L4 are the width and length of the transistors M4 and M5
W6 and L4 are the width and length of the transistors M6 and M7
W8 and L8 are the width and length of the transistors M8 and M9
W10 and L10 are the width and length of the transistors M10 and M11

augmented 40 times (almost 700 K samples) for the same 5000 epochs. Its training took approximately 8 h. ANN-3 was also trained on the same augmented dataset but only 500 epochs, and it was initialized with weights from ANN-1. Its training took less than an hour. Their performance after training on the training and validation sets is summarized in Table 4.4.

In terms of performance, ANN-2 and ANN-3 showed the best results. From Table 4.4, we can observe that MSE and MAE error for the training and validation sets were lower for these networks when compared to ANN-1, but not by a long margin. In terms of individual MAE for each device's sizes, results were very similar across all ANNs, with a slight advantage for ANN-1. Table 4.5 shows the MAE between the predicted and true devices' sizes for each design variable of the designs in the test set.

Using the ANN for Circuit Sizing

To emulate real-world usage of the models, ANNs were used to predict 100 random samples with a deviation of up to 15% from the specifications, as described in Sect. 4.3.2. The predicted designs were then simulated using the circuit simulator. Selection of the best solution was done by FoM for target 1, GBW for target 2, and IDD and FoM for target 3. It is easily seen by the performance of the obtained circuits that the ANNs learned the design patterns and can even extrapolate for specifications outside those of the training data. Moreover, circuits with FoMs larger than 1000

Table 4.6 Performance of sampled designs for the VCOTA

	#HF[a]	DC Gain (dB)	GBW (MHz)	IDD (μA)	PM (°)	FoM[b]
Target 1		50	60	300	65	
ANN-1	0.33	50	63	318	64	1180
ANN-2	1	51	61	320	65	1153
ANN-3	1	51	63	325	65	1165
Target 2		40	150	700	55	
ANN-1	0.24	44	148	822	54	1082
ANN-2	0.21	49	60	325	73	1106
ANN-3	1	43	100	509	61	1182
Target 3		50	30	150	65	
ANN-1[c]	0.73	50	3	141	74	116
ANN-1		50	30	205	74	889
ANN-1[d]		49	67	329	60	1215
ANN-2[c]	1	54	38	240	71	950
ANN-2[d]		54	46	268	64	1033
ANN-3[c]	0.97	55	30	217	69	842
ANN-3[d]		54	54	309	56	1050

[a]Ratio of the number of solutions with FoM higher than 850 (the min value in the training data was 900) to the total number of samples; [b]MHz.pF/mA; [c]Best IDD; [d]Best FoM

were obtained in all predictions of the ANNs. The FoM used in this example, which was an optimization target in the process used to obtain the dataset, is defined as:

$$\text{FoM} = \frac{\text{GBW} \times C_{\text{Load}}}{\text{IDD}} \tag{4.10}$$

Table 4.6 summarizes the performance of the best circuits predicted by ANN1, ANN2, and ANN3 for three target specifications. In this experiment, ANN-1 seems to be more flexible when asked to predict the sizing for new specifications; on the other hand, it produces many unfeasible designs when sampled. ANN-2 and ANN-3 are more stable and both always generate good designs when sampled inside the training data. ANN-2 shows more limitations when trying to explore new specifications. ANN-3, because it used transfer learning from ANN-1, is more flexible to new specifications, but still lags when compared to ANN-1.

4.5.2   Two-Stage Miller Amplifier

For the two-stage Miller amplifier, the same approach was followed. The structure considered also has 15 input variables, obtained from the second-order polynomial

Table 4.7 Performance of the ANNs trained for the two-stage amplifier

		MSE train	MSE Val.	MAE Train	MAE Val.
Two-stage Miller	ANN-4	0.0561	0.0414	0.1357	0.0937
	ANN-5	0.0072	0.0073	0.0590	0.0595
	ANN-6	0.0072	0.0073	0.0590	0.0597

Table 4.8 Average MAE of the devices' sizes for the two-stage amplifier

Design variables	ANN-4	ANN-5	ANN-6	Design variables	ANN-4	ANN-5	ANN-6
Wb	1.27E−05	2.75E−06	2.86E−06	NFbp	4.27E−00	5.64E−01	5.89E−01
Wp	7.76E−05	8.59E−06	8.85E−06	NFb2	8.91E−00	6.49E−01	6.55E−01
Wal	2.78E−05	4.01E−06	4.20E−06	NFal	9.34E+01	1.69E+01	1.77E+01
W2g	7.35E−05	8.76E−06	8.86E−06	NFp	1.35E+02	2.84E+01	2.86E+01
Lb	4.65E−06	7.77E−07	8.15E−07	NF2 g	1.30E+02	2.84E+01	2.84E+01
Lp	4.54E−06	6.06E−07	6.11E−07	Lc	7.49E−05	1.18E−05	1.20E−05
Lal	8.09E−06	9.41E−07	9.47E−07	NFc	1.44E+02	2.30E+01	2.31E+01
L_2g	5.43E−06	6.57E−07	6.67E−07				

feature extension of the same 4 performance figures, 3 hidden layers with 120, 240, 60 nodes each, and the output layer has 15 nodes. Again, three ANNs were trained. ANN-4 was trained on the original dataset, for 5000 epochs with batches of 512 samples, taking approximately 12 min to conclude the training. ANN-5 was trained on the dataset augmented 20 times (more than 100 K samples) for 500 epochs. Its training took approximately 20 min. The training set was comprised of 96% of total data, while the validation set of 4%. ANN-6, was trained on the dataset augmented 20 times, for 500 epochs with batches of 512 samples, initialized with weights from ANN-4. The training set and the validation set were split with a 50% ratio (since the dataset was augmented, there is no problem in choosing the same percentage for both sets). Its training took approximately 20 min. Their performance after training on the training and validation sets is summarized in Table 4.7.

In terms of performance, ANN-5 and ANN-6 showed the best results. From Table 4.7, we can observe that MSE and MAE for the training and validation sets were significantly lower for these networks, when compared to ANN-4. Table 4.8 indicates the average MAE between all the predicted and true devices' sizes from the test set. We can see that ANN-4 performed much worse than ANN-5 and ANN-6, yielding higher error than almost all design variables.

Using the ANN for Circuit Sizing

In Table 4.9, we see that ANN-5 and ANN-6 can generate stable designs, despite being inaccurate for some design variables. ANN-4 shows some limitations for one target but can generate designs similar to the other networks for the remaining targets.

Table 4.9 Performance of sampled designs for the two-stage amplifier

	DC gain (dB)	GBW (MHz)	IDD (μA)	PM ($^\circ$)
Target 1	70	10	30	65
ANN-4	77	9	39	65
ANN-5	78	9	35	64
ANN-6	77	8	36	64
Target 2[a]	100	10	30	65
ANN-4	95	12	48	89
ANN-5	87	6	41	68
ANN-6	85	13	39	42
Target 3[a]	70	2	10	65
ANN-4	80	3	29	70
ANN-5	75	1	19	75
ANN-6	72	1	18	75

[a]Targets outside the performances present in the dataset

4.5.3   Test Case—Classification and Regression Model

For the multicircuit classification and regression model, the two previously studied circuits were again considered. The goal of this architecture is not only to learn design patterns from these circuits, but also to identify which circuit should be considered to implement the target specifications. To do this, regression is applied to learn devices' sizes and classification is used to learn circuit classes.

For this example, the used dataset, i.e., Dataset-3, has 15,000 different design points. Each third of the dataset belongs to three different chunks of data: two classes of circuits and one additional group of data. The first class refers to circuit specifications that belong to the VCOTA topology (encoded as 001); the second class refers to circuit specifications that belong to the two-stage Miller topology (encoded as 010); the additional group of data is comprised of augmented data built up from the other two circuits (encoded as 100), but designed to not meet any of the specifications required by those circuits (i.e., maximum and minimum performance specifications are outside the required ranges). This last chunk was added to the problem to check if the network would be misled by input specifications that are out of the ranges required for the two studied circuits. The circuit performances that were considered to train the ANN were again DC gain, IDD, GBW, and PM, and the ranges of values found in the dataset are given in Table 4.10.

ANN Structure and Training

Three ANNs were trained for this dataset, i.e., ANN-7 to ANN-9. The structure considered has 15 input variables (obtained from the second-order polynomial feature extension of the 4 performance figures from Table 4.10), 3 hidden layers with 120,

Table 4.10 Performance ranges in the combined dataset

		DC gain (dB)	GBW (MHz)	IDD (μA)	PM (°)
VCOTA	Max	56.8	78	395	80
	Min	44.7	34	221	60
Two-stage Miller	Max	97.2	102.8	0.8	89.9
	Min	59.8	1.5	0.3	55
Augmented data	Max	117.1	33.2	0.3	89.9
	Min	69.7	1.5	0.1	55

Table 4.11 Performance of the classification and regression ANNs

	Loss		Regression (MAE)		Classification (SSCE)	
	Train	Val.	Train	Val.	Train	Val.
ANN-7	0.0033	0.0034	0.0324	0.0329	1.0	1.0
ANN-8	0.0033	0.0033	0.0323	0.0324	0.9999	0.9999
ANN-9	0.0033	0.0033	0.0322	0.0321	0.9999	0.9998

240, 60 nodes each, and the output layer has 30 nodes, which represent the different devices' sizes of the VCOTA and two-stage Miller topologies (12 and 15 nodes, respectively), and the class to which they belong to (3 nodes to encode each of the three classes). For this dataset, the augmented data does not have any devices' sizes specified. Only performance figures were specified for this chunk of the dataset (as input features) so that a different class of circuits could be simulated.

ANN-7 was trained on the original dataset, for 5000 epochs with batches of 512 samples. Its training took less than 46 min. ANN-8 was trained on the augmented dataset, where 75 K samples were generated for each circuit class, but only for 500 epochs. The network was initialized with weights from ANN-7. Its training took less than 50 min. ANN-9 was trained on the augmented dataset, where 100 K samples were generated for each circuit class, for 5000 epochs. Its training took approximately 12 h. Three performance metrics were used: the custom loss function from 3.8, MAE for the regression nodes, and SSCE for the classification nodes. Their performance after training on the training and validation sets is summarized in Table 4.11. Table 4.12 indicates the average MAE between all the predicted and true devices' sizes, and class prediction accuracy. In terms of performance, all three ANNs showed favorable results.

From Table 4.11, we can observe that loss, regression, and classification errors were similar for all three networks. The MAE for each individual device's sizes was also very similar across all ANNs.

Table 4.12 MAE of the devices' sizes and class prediction accuracy

Circuit	Design variables	ANN-7		ANN-8		ANN-9	
		MAE	Class Acc. (%)	MAE	Class Acc. (%)	MAE	Class Acc. (%)
VCOTA	w8	7.51E−09	100	6.64E−09	100	6.08E−09	100
	w6	5.78E−06		5.82E−06		5.69E−06	
	w4	2.38E−06		2.19E−06		2.21E−06	
	w10	1.68E−06		2.22E−06		1.50E−06	
	w1	1.27E−06		9.09E−07		8.71E−07	
	w0	6.73E−06		6.92E−06		7.30E−06	
	l8	1.71E−08		1.75E−08		1.54E−08	
	l6	5.59E−08		5.64E−08		5.48E−08	
	l4	5.13E−08		4.93E−08		4.94E−08	
	l10	2.89E−08		2.76E−08		2.63E−08	
	l1	4.26E−08		4.14E−08		3.89E−08	
	l0	2.67E−08		2.45E−08		2.84E−08	
Two-stage Miller	wb	1.18E−06	100	1.03E−06	100	1.05E−06	100
	wp	4.00E−06		3.64E−06		4.21E−06	
	wal	1.57E−06		1.72E−06		1.65E−06	
	w2 g	3.34E−06		3.18E−06		3.82E−06	
	lb	3.37E−07		3.05E−07		3.20E−07	
	lp	2.31E−07		2.58E−07		2.68E−07	
	lal	4.08E−07		3.82E−07		3.89E−07	
	l2g	2.79E−07		2.47E−07		2.82E−07	
	mbp	2.76E−01		2.64E−01		2.77E−01	
	mb2	3.73E−01		3.39E−01		3.65E−01	
	_mal	6.75E+00		6.98E+00		6.72E+00	
	mp	1.28E+01		1.00E+01		1.18E+01	
	m2 g	1.14E+01		1.10E+01		1.14E+01	
	lc	4.65E−06		4.41E−06		4.78E−06	
	nfc	9.44E+00		8.73E+00		8.87E+00	

Using the ANN to Predict New Designs

This model is more complex and can be trained on multiple topologies for the same analog function and selects the most appropriate solution and its sizing. In Table 4.13, we can see that ANN-9 can generate stable solutions for either topology, despite the considerable variability verified in some design variables.

Table 4.13 Performance of sampled designs

	DC gain (dB)	GBW (MHz)	IDD (μA)	PM (°)	Topology
Target 1	50	50	300	60	
ANN-9	54	58	322	57	VCOTA
	54	58	323	58	
	54	58	322	58	
Target 2	70	10	70	60	
ANN-9	83	10	64	59	Two-stage Miller
	83	10	63	59	
	83	10	62	59	

4.6 Conclusions

In this chapter, the design flow for this work was presented, as well as two ANN architectures: a regression-only model and a classification and regression model. For both models, their structure and hyper-parameter selection were explained, as well as the process of polynomial features and data normalization applied to the input training data. The selection of the correct hyper-parameters, as being one of the most important tasks during the preparation of the model, was discussed. Furthermore, the preparation of the data for the training phase of the model was also analyzed. In particular, the concept of polynomial features was introduced. This is a good way to give the model more flexibility by overcoming the problem of having a low amount of input features.

Details regarding the training phase of the models, the concept of transfer learning and sampling results from the ANNs were also addressed. The use of L_2 regularization was proposed as a way to avoid overfitting of the models. Transfer learning was introduced as being a popular practice in the deep learning community, where ANNs trained with large datasets are repurposed for problems with fewer available data points. Finally, the process of sampling, where outputs of the ANNs are filtered to avoid possible biased solutions, was discussed.

The results for the two architectures were presented. In the regression-only model, three ANNs were tested for each circuit topology, VCOTA and two-stage Miller. In the regression and classification model, three ANNs were tested for a dataset comprised of design points from both topologies.

For both architectures, low error on the training and validation sets was achieved. In terms of prediction accuracy rate, the results were more favorable for the VCOTA topology in both architectures, where a lower prediction error was achieved. The ANNs for the two-stage Miller amplifier performed was not as good, but still both models were able to predict reasonable circuits when used to predict new designs.

References

1. Cadence, Virtuoso Analog Design Environment GXL [Online]. Available: http://www.cadence. com. Accessed 15 May 2019
2. MunEDA, WIKED™ [Online]. Available: http://www.muneda.com. Accessed am 15 May 2019
3. R. Martins, N. Lourenço, N. Horta, J. Yin, P. Mak, R. Martins, Many-objective sizing optimization of a class-C/D VCO for ultralow-power IoT and ultralow-phase-noise cellular applications. IEEE Trans. Very Large Scale Integr. (VLSI) Syst. **27**(1), 69–82 (2019)
4. F. Passos et al., A multilevel bottom-up optimization methodology for the automated synthesis of RF systems. IEEE Trans. Comput. Des. Integr. Circuits Syst. (2019). https://doi.org/10.1109/ TCAD.2018.2890528
5. A. Canelas, et al., FUZYE: a fuzzy C-means analog IC yield optimization using evolutionary-based algorithms. IEEE Trans. Comput. Aided Des. Integr. Circ. Syst. (2018). https://doi.org/ 10.1109/tcad.2018.2883978
6. R. Gonzalez-Echevarria et al., An automated design methodology of RF circuits by using pareto-optimal fronts of EM-simulated inductors. IEEE Trans. Comput. Des. Integr. Circ. Syst. **36**(1), 15–26 (2017)
7. R. Martins, N. Lourenço, F. Passos, R. Póvoa, A. Canelas, E. Roca, R. Castro-López, J. Sieiro, F.V. Fernández, N. Horta, Two-step RF IC block synthesis with pre-optimized inductors and full layout generation in-the-loop. IEEE Trans. Comput. Aided Des. Integr. Circuits Syst. (TCAD) **38**(6), 989–1002 (2019)
8. G. Wolfe, R. Vemuri, Extraction and use of neural network models in automated synthesis of operational amplifiers. IEEE Trans. Comput. Aided Des. Integr. Circ. Syst. **22**(2), 198–212 (2003)
9. H. Liu, A. Singhee, R.A. Rutenbar, L.R. Carley, Remembrance of circuits past: macromodeling by data mining in large analog design spaces, in *Proceedings 2002 Design Automation Conference* (2002). https://doi.org/10.1109/dac.2002.1012665
10. N. Kahraman, T. Yildirim, Technology independent circuit sizing for fundamental analog circuits using artificial neural networks, in *Ph.D. Research in Microelectronics and Electronics* (2008)
11. I. Goodfellow, Y. Bengio, A. Courville, *Deep Learning*. MIT Press (2016)
12. N. Lourenço, R. Martins, N. Horta, *Automatic analog IC sizing and optimization constrained with PVT corners and layout effects*. Springer International Publishing (2016)
13. V. Nair, G.E. Hinton, Rectified linear units improve restricted boltzmann machines, in *Proceedings of the 27th International Conference on International Conference on Machine Learning* (2010)
14. D.P. Kingma, J. Ba, Adam: a method for stochastic optimization, in *Proceedings International Conference on Learning Representations (ICLR)* (2015)
15. R. Póvoa, N. Lourenço, R. Martins, A. Canelas, N. Horta, J. Goes, Single-stage amplifier biased by voltage-combiners with gain and energy-efficiency enhancement. IEEE Trans. Circ. Syst. II Express Briefs **65**(3), 266–270 (2018)
16. M. Abadi, et al. TensorFlow: large-scale machine learning on heterogeneous systems, 2015. Software available from tensorflow.org

Chapter 5
ANNs as an Alternative for Automatic Analog IC Placement

5.1 Layout Synthesis by Deep Learning

In the literature, automatic layout generation is usually solved using one of three different methodologies, which are distinguished by the amount of legacy knowledge that is used during the generation: *constrained placement generation* [1–7], *constrained routing generation* [8–10] or complete *constrained place and route* [11–13] solutions (i.e., without considering legacy data), *layout migration or retargeting* [14–16] from a previous legacy design or template, and *layout synthesis with knowledge mining* [17, 18] from multiple previously designed layouts. While none of them actually succeeded in industrial applications, recent developments in artificial intelligence and machine learning (ML) and, their rising success in different fields of study, indicate that the knowledge mining route might bring promising results for the automation attempts of this task.

In this chapter, exploratory research using ML, and more specifically, ANNs, is conducted for analog IC placement. Similar to the *layout migration or retargeting* from a previous legacy design or template and *layout synthesis with knowledge mining*, the methodology abstracts the need to explicitly deal with the high amount of constraints from the *layout generation considering placement and routing constraints* by learning patterns from validated legacy designs. These should contain all constraints considered relevant by the layout designer or EDA engineer. The proposed model extends the knowledge mining of the *layout synthesis with knowledge mining* [17, 18] as it is an end-to-end approach without the need to define any kind of tiebreaker manually. More, the produced mapping embeds reusable design patterns that generalize well beyond the training data and provide promising layout solutions for new designs. Moreover, it is intended to demonstrate how ANNs can be interesting tools to output different placement alternatives for the same circuit sizing at push-button speed, e.g., placements with different aspect ratios.

© The Author(s), under exclusive license to Springer Nature Switzerland AG 2020   67
J. P. S. Rosa et al., *Using Artificial Neural Networks for Analog
Integrated Circuit Design Automation*, SpringerBriefs in Applied
Sciences and Technology, https://doi.org/10.1007/978-3-030-35743-6_5

5.2 Development of an ANN Model

In this subsection, the circuit used to develop and test the ML system is introduced, and then, the specific architectures of the dataset and of the ANN are described. The following intrasections discuss the methodologies and techniques used to increase the performance of the ANN for this specific problem. In order to compare models and take some conclusions on which is more suited, a set of metrics to evaluate their performance and its respective advantages are introduced.

5.2.1 Circuit Used for Tests

The circuit chosen to demonstrate the development of the ANN model for analog IC placement is the single-stage amplifier with voltage combiners proposed in [19], whose schematic is shown in Fig. 5.1, with the current flows from V_{DD} to V_{SS} highlighted. This circuit was chosen due to its moderate complexity, as it contains twelve devices grouped into six symmetry pairs, whose layout locations must be determined. As characteristic from analog IC placement, there are an uncountable number of different possible placement solutions for this circuit, even when considering the same circuit sizing. These different designers' preferences can be represented as specification-independent templates that encode a set of topological relations between devices. These topological relations between cells are kept as the sizing of devices is changed within a broad range of possible values. Examples of two different placement templates for the single-stage amplifier using voltage combiners are represented in Fig. 5.2 with the current flows highlighted.

Fig. 5.1 Single-stage amplifier using voltage combiner's schematic [19]

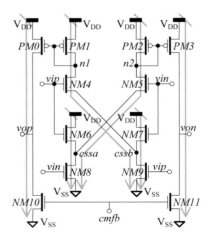

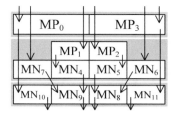 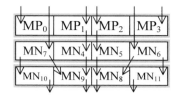

Fig. 5.2 Two different placement templates with current flows highlighted that encode a set of topological relations between devices [20, 21]

5.2.2 Dataset Architecture

The sizing of the devices that is used as input to the ANN is the transistors' total width, transistors' total length, transistors' number of fingers, and also the total width and height of the parametric cell of that device, i.e., the real width and height measurements of the layout implementation of each device. The width and height of the devices' layout are obvious choices as they define the real area occupied by it, even though the remaining parameters (transistors' total width, length, and number of fingers) define its physical properties.

As stated, a unique sizing solution for the devices can produce an uncountable number of different and valid placement solutions. To take this into account for real-world extrapolation, the dataset used contains placement solutions for twelve different templates for the same sizing solutions that encode twelve topological relations between cells, simulating the layout-style preferences of twelve different designers. An example of different placement solutions generated from three different templates for the same sizing is shown in Fig. 5.3. Even though twelve different templates are available, the dataset used to train the ANN is set to contain only three different placement solutions, i.e., the placement with the smallest area, the placement with the smallest height (designated by maximum aspect ratio), and the placement with the smallest width (designated by minimum aspect ratio). The information of the index of the template that produced that layout is kept for future consideration. As the original dataset (with twelve templates) contains the area, total height, and the total width of each placement solution, it is easily processed to generate a new dataset where for each sizing solution only the placement with the smallest area, width and height are contained. The original dataset for the single-stage amplifier circuit using voltage combiners contains around 10,000 samples, where 80% were used to train the network and 20% used to test the dataset.

A section of the dataset is represented in Tables 5.1 and 5.2. In the first table, the columns that correspond to the inputs of the model are represented, where the transistors' total width, length, and number of fingers corresponding to the columns starting with *w, l,* and *nf,* respectively, and, *wt* and *ht,* to the width and height of the physical implementation of the device, respectively. The second table contains some columns with the layout information (placement's *width, height, area,* and

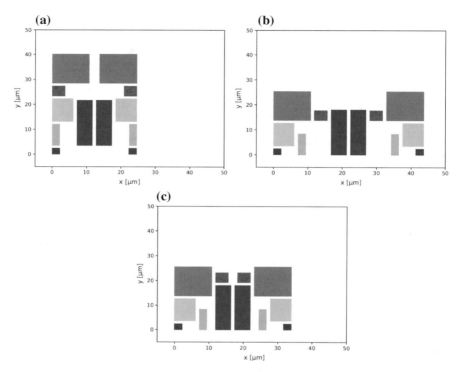

Fig. 5.3 Different layout placement solutions, generated from three different templates, for the same sizing

coordinates of each cell) produced by each template. Finally, the remaining columns starting with $x\_$ and $y\_$ contain the coordinates of the bottom left corner of each device in the corresponding template. As previously explained, the dataset used to train the ANN to follow the same rules but only contains three templates instead of Template$_1$ to Template$_{12}$.

Feature engineering is an essential part of the creation of an ML model, where modifications on the features of the model or elimination of some features that are not important are performed, so better results are achieved with the same dataset. Adding polynomial features is an example of this. This process increases the number of inputs by generating new features that correspond to the polynomial built of existing features. For example, if considering features a, b, and c, the polynomial features of the second degree are: a, b, c, a^2, $a.b$, $a.c$, b^2, $b.c$, and c^2. This increase in the number of inputs opens the possibility of making the model to learn new patterns that are only present with polynomial combinations of the input variables. For example, in this work it is important to have the area of the physical implementation of each device area as an input, which is easily accomplished by the adding of polynomial features, corresponding to the multiplication of the device's width by the device's height features.

Table 5.1 Section of the input columns belonging to the original dataset with all templates

Device$_0$					...	Device$_{11}$				
w$_0$ (μm)	l$_0$ (μm)	nf$_0$ (μm)	wt$_0$ (μm)	ht$_0$ (μm)	...	w$_{11}$ (μm)	l$_{11}$ (μm)	nf$_{11}$ (μm)	wt$_{11}$ (μm)	ht$_{11}$ (μm)
17.5	0.36	3	3.48	8.07	...	8.7	0.89	3	4.73	5.14
17.7	0.36	7	6.52	4.765	...	83.3	0.83	7	9.47	14.14
50.7	0.40	1	2.0	52.06	...	20.8	0.77	3	4.37	9.17

Table 5.2 Section of the column with the placement information belonging to the original dataset with all templates

Template$_1$								...
Width (μm)	Height (μm)	Area (μm^2)	x_0 (μm)	y_0 (μm)	...	x_11 (μm)	y_11 (μm)	...
20.17	101.77	2052.7	10.71	4.01	...	15.44	4.01	...
35.73	98.79	3529.8	18.49	4.01	...	26.26	4.01	...
23.15	109.0	2523.5	12.2	3.99	...	18.78	3.99	...

...	Template$_{12}$							
...	Width (μm)	Height (μm)	Area (μm^2)	x_0 (μm)	y_0 (μm)	...	x_11 (μm)	y_11 (μm)
...	24.23	60.46	1464.9	6.69	6.39	...	6.69	0.0
...	33.57	59.69	2003.7	6.69	15.39	...	6.69	0.0
...	28.79	132.79	3823.0	9.4	10.42	...	9.4	0.0

5.2.3 Neural Network Architecture

This chapter proposes the development of a model that predicts the placement of all the devices of an analog circuit based solemnly on their sizing. The proposed model is a nonlinear model described by an ANN. To achieve this purpose, the inputs of the model are some measures of the size of the devices and the respective output must be the actual position of the devices on the layout. Since the number of inputs and outputs vary with the number of devices in the circuit, at this stage of the development, each model can only describe a specific circuit or, at most, circuits with the same number of devices. In order to offer a broader layout possibilities of placement solutions to the designer, the output of the ANN is triplicated so that the output is three distinct placement solutions: the one with the smallest area and the ones with the smallest and largest aspect ratio, in case the designer needs different aspect ratios to fit his system-level layout. Layout placements for these three categories are illustrated in Fig. 5.4 for the same circuit sizing. A general network that represents the described architecture is represented in Fig. 5.5. The number of neurons in each layer is variable, as well as the number of inputs layers.

5.2.4 Preprocessing the Data

Preprocessing the dataset is a fundamental step of any ML process, and it is strongly dependent on the problem and on what the dataset represents. In this chapter, the dataset contains sizes of devices and their respective location. Since the electronic devices of nanometric integration technologies present extremely small dimensions, they also result in extremely small layouts. This fact causes the values present in the

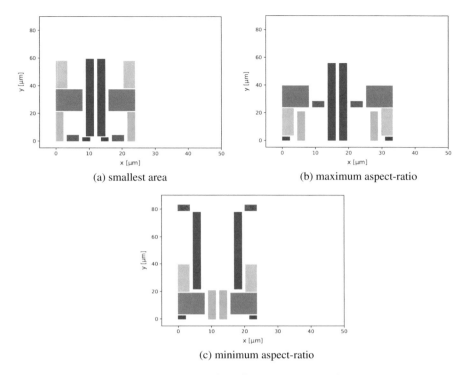

(a) smallest area (b) maximum aspect-ratio

(c) minimum aspect-ratio

Fig. 5.4 Placement for the same sizing, but for different layout categories

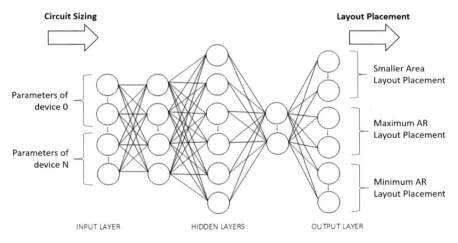

Fig. 5.5 Architecture used to predict layout placements given the sizing

dataset to be in the order of the nanometers or micrometers and therefore will make the gradients computed in the backpropagation method of the ANN to be even smaller, to a point where a computer might not be able to represent them, and consequently, the updates on the weights of the network will be performed incorrectly. To solve this problem, it is a common procedure to scale the dataset so that the mean of each column of the dataset is 0 and the variance is unitary. This is done using Eqs. 5.1, 5.2, and 5.3, where z is the scaled column of the dataset, x is the unscaled value of the same column, μ is its mean, and σ its variance. There is also the possibility of scaling each column of the dataset so that its values are set between 0 and 1, following Eq. 5.4, where x_{min} and x_{max} represent the minimum and the maximum value of the unscaled column. This scaling method sets the maximum value to 1 and the minimum value to 0, and every value in between them to be proportionally scaled between 0 and 1.

$$y = \frac{x - \mu}{\sigma} \tag{5.1}$$

$$\mu = \frac{1}{N} \sum_{i=1}^{N} x_i \tag{5.2}$$

$$\sigma = \sqrt{\frac{1}{N} \sum_{i=1}^{N} (x_i - \mu)^2} \tag{5.3}$$

$$z = \frac{x - x_{min}}{x_{max} - x_{min}} \tag{5.4}$$

Besides scaling, it is also a good practice to transform the data in some way that makes it easier for the algorithm to learn patterns in the problem. The proposed outputs of the model are the coordinates of the bottom left corner of each device, and often it is assumed that the bottom left corner of the layout is located at $(0, 0)$ coordinate. However, since analog IC layout is usually symmetric, it is possible to assume that the symmetry axis is located at $x = 0$ and therefore make the output of the model to be symmetrical values that represent the position of the devices, such as the center point of each device or a left corner for the left-side devices and the right corner for the right-side devices, as illustrated in Fig. 5.6.

5.2.5 Metrics to Evaluate the Models

In order to evaluate a trained model, some metrics must be developed. The model is evaluated using a different set (previously denominated test set) than the one used to train the model, being that data something that the model has never seen. The ultimate objective is always to extrapolate how could the model generalize the acquired knowledge and behave in a real-world application.

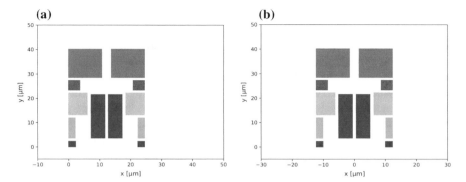

Fig. 5.6 Placement representation **a** before and **b** after centering the data

5.2.5.1 Error-Related Metrics

The most evident metrics to be used are error metrics, such as the mean squared error (MSE), the mean absolute error (MAE), or the mean absolute error per device (MAED). These metrics measure how much the placement predicted by the model differs from the target placement present in the dataset and can be computed as shown in Eqs. 5.5, 5.6, and 5.7, respectively, where N is the number of examples in the test set, M is the number of devices in the circuit, x^p is the predicted x-coordinate of device j in example i, and x^t is the target x-coordinate for the same case. The MSE is used to train the model since it is easily differentiable; however, it is not a good measure to represent a real error since it corresponds to a squared error. The MAE represents the error of the placement across all devices (in the units of the output) that is expected when the model predicts a placement. The MAED indicates how much the placement of a device differs from the target placement on average. Since both the MAE and the MAED represent an error in the dimensions of the output variables, they are easier to interpret metrics.

$$\text{MSE} = \frac{1}{N} \sum_{i=0}^{N} \sum_{j=0}^{M} \left[\left(x_{ij}^p - x_{ij}^t \right)^2 + \left(y_{ij}^p - y_{ij}^t \right)^2 \right] \tag{5.5}$$

$$\text{MAE} = \frac{1}{N} \sum_{i=0}^{N} \sum_{j=0}^{M} \sqrt{\left[\left(x_{ij}^p - x_{ij}^t \right)^2 + \left(y_{ij}^p - y_{ij}^t \right)^2 \right]} \tag{5.6}$$

$$\text{MAED} = \frac{1}{N * M} \sum_{i=0}^{N} \sum_{j=0}^{M} \sqrt{\left[\left(x_{ij}^p - x_{ij}^t \right)^2 + \left(y_{ij}^p - y_{ij}^t \right)^2 \right]} \tag{5.7}$$

5.2.5.2 Overlap-Related Metrics

Although computing the error of a prediction is a valuable metric, in this specific problem of analog IC placement, if the error of a prediction is large, it does not necessarily mean that the predicted placement is wrong, since it can still be a valid placement solutions and it might even be a better placement alternative than the target placement. To evaluate these cases, there is a need to introduce a metric that evaluates if the circuit is valid or not. Following, it is computed the total overlap area among all the devices, as the model might output a placement where this situation may occur, and it is a clear indicator of an invalid placement. Computing the total overlap area for each example turns it possible to compute the mean overlap area (MOA) over a set of examples.

5.2.5.3 Accuracy

Since this problem is in its core a regression problem, there is no direct way of knowing if the model output is correct or not. To have a metric representing the accuracy of the model, it was considered that the output can be assumed "correct" if the error or the overlap was below a reasonable threshold. It is acknowledged that overlaps below that threshold could be simply eliminated by a deterministic post-processing algorithm. These new metrics are used to compute the percentage of cases where the model gave an output in which the error or the amount of overlap can be considered "correct", i.e., error accuracy (EA) and overlap accuracy (OA). To count the cases where both the error and the overlap of the output placement were minimal, it is possible to compute the error and overlap accuracy (EOA).

5.2.6 Training of the Network

To train the ANN and build one that has the optimal performance, it is necessary to tune several different hyper-parameters. This is one of the most difficult tasks on these types of models, as it is an iterative and time-consuming process to train different networks to compare its performances. In order to turn this process easier, in this work, some hyper-parameters were set to default values as they proved to achieve good performances in several test cases. These were the case of the activation function and the hyper-parameters corresponding to the optimizer used to find the minima of the cost function. The activation function in the hidden layers was set to the ELU function [22], and the linear function was used in the output layer. The optimizer used was the Adam optimizer [23], and its parameters were set to the default values: $\alpha = 0.001$, $\beta 1 = 0.9$, and $\beta 2 = 0.999$.

The remaining hyper-parameters were set by trial and error, and the batch size was set to 500 as it is a moderate value by the time of writing. Since the training dataset contains around 8000 examples, 500 is around 6.25% of the dataset which is lower than the 10% that is considered a large batch size. The downside of this lower value is that the training times become longer while requiring fewer epochs to achieve convergence. The number of epochs and the number of neurons were set by trial and error by training the different networks and comparing their performances. The results of these tests are shown in Sect. 5.3 of this chapter.

The search for the ANN architecture that achieves the best performance will be an iterative process. First, a network with no hidden layers will be tested to assess if linear regression is enough to provide satisfactory results. The hyper-parameters of the network will be tuned to check its effect on the performance. After, polynomial features will be added to verify its impact on the performance of the model. Afterward, simpler networks with few hidden layers will be tested first, and then, their depth will be gradually increased until the performance is acceptable or until the training times and/or memory needed become problematic. Moreover, since centering the data might make it easier for the algorithm to learn the patters on the dataset, it will be applied and its impact studied.

5.3 Experimental Results

The ANNs were implemented in Python language due to the availability of several packages designed for this task such as TensorFlow [24] and Scikit-Learn [25]. Pandas package [26] was used to process the dataset and Matplotlib [27] was used to plot the outputs of the model. The networks were trained on an Intel®Core™i7 processor with 8 GB of RAM. The thresholds used to verify the correctness of the placement solutions generated by the model were 500 nm for EA and 0.05 μm^2 for OA.

5.3.1 Single-Layer Perceptron

Before entering deeper architectures, a simpler approach was evaluated with a network built with only an input layer and an output layer, called single-layer perceptron.

5.3.1.1 Linear Regression

This ANN architecture represents a linear regression, where the outputs, y, are given by a function of an array of constants, W, multiplied by the input variables, x. The array of constants corresponds to the weights that are trained for the ANN. The first objective of this test was to evaluate if a single-layer network could present acceptable results without preprocessing the data (only scaling was applied). The results of this experiment are shown in Table 5.3, where it is possible to observe that between 1500

Table 5.3 Performance of trained single-layer networks

Epochs	Test					Train				
	MAED (nm)	MOA (nm)	EA (%)	OA (%)	EOA (%)	MAED (nm)	MOA (nm)	EA (%)	OA (%)	EOA (%)
250	548.071	0.182	60.701	98.081	60.701	532.361	0.089	60.602	97.925	60.590
500	540.764	0.175	61.372	97.937	61.372	525.100	0.083	61.178	97.961	61.178
1000	545.149	0.145	61.756	97.937	61.756	524.496	0.092	61.334	97.913	61.322
1500	542.229	0.140	61.612	97.937	61.612	524.244	0.087	61.298	97.973	61.298
2000	549.320	0.127	61.084	97.937	61.084	530.044	0.091	60.686	97.949	60.662
3000	549.122	0.137	61.180	97.937	61.180	528.296	0.093	60.854	97.949	60.842

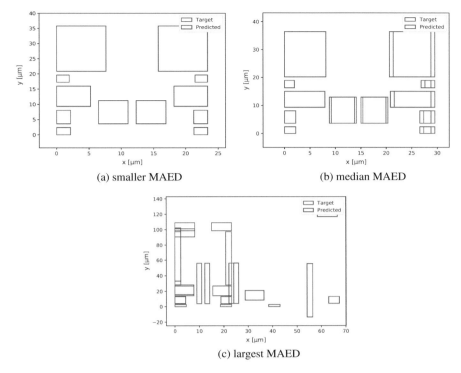

Fig. 5.7 Target and predicted placement for some examples of the testing set, using the single-layer network trained with 2000 epochs

and 2000 epochs the MAED on the test set starts to increase, which is a clear sign of overfitting. The linear regression model presents approximately an EOA of 60%, and further developments are needed to improve this value. Nonetheless, this initial test indicates that the patterns present in the dataset can be learned by a network.

To illustrate the performance of the model, three outputs for different sizing solutions of the single-layer network trained with 2000 epochs are presented in Fig. 5.7, where (a) corresponds to the sizing with smallest MAED error, (b) to the median MAED error, and (c) to the maximum MAED error. While the worst case cannot be considered a valid placement, the median and best case are close to the target placement. For the next cases, the example with the lowest MAED is omitted, as the predicted placement solution is virtually identical to the target.

5.3.1.2 Polynomial Features

In order to prove that this problem can be modeled using a network, in the next experiment the dimension of the input variables is augmented using polynomial features to force overfitting of the model. The model was trained during 2000 epochs

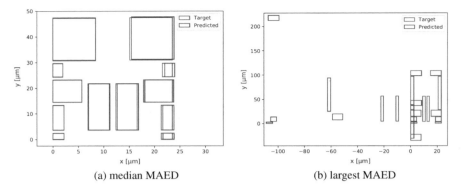

(a) median MAED   (b) largest MAED

Fig. 5.8 Target and predicted placement for some examples of the testing set, using the single-layer network with second-order polynomial features

since it was around this number that the single-layer model started to overfit. The results are shown in Table 5.4. The introduction of the polynomial features resulted in an increase in the accuracy of the model and a reduction of the MAED. Due to the larger input space, not only the MAED in the training set was decreased but also in the test set. Although the MAED decreases and the EA increases, both MOA and OA also increase in the training and the test set. This is an indicator that the model is overfit to some specific examples that might have similar inputs. By observing the training curves of Fig. 5.9, it is possible to identify the moment the model starts to overfit to the training data, around 500 epochs, as the MSE in the testing set starts to increase. In this regard, in the next tests, a regularization technique used is early stopping by setting the number of epochs to a value that leads to a non-overfit model. The predicted placements that correspond to the median and largest error are presented in Fig. 5.8. It can be concluded that this network architecture has its limitation as it cannot predict a valid layout placement for a large number of circuit sizing solutions. Polynomial features of third degree and higher were not considered due to the enormous computational power and memory required, as the number of weights increases exponentially too.

5.3.2   Multilayer Perceptron

In the search for a model with better performance, it is necessary to deepen the network by increasing the number of hidden layers. This will raise also the need to verify the effect of both the depth of the ANN and the number of units in each hidden layer. These networks will be trained during 1000 epochs due to their increased complexity. Since the introduction of polynomial features brought positive impact in performance, the tests with deeper networks will also consider second-degree polynomial features.

Table 5.4 Trained single-layer networks with polynomial features

Degree	N inputs	Test						Train					
		MAED (nm)	MOA (nm)	EA (%)	OA (%)	EOA (%)		MAED (nm)	MOA (nm)	EA (%)	OA (%)	EOA (%)	
–	60	542.327	0.133	61.564	97.937	61.564		524.221	0.087	61.250	97.973	61.250	
2	1831	392.856	0.184	80.614	98.464	80.614		348.472	0.103	81.794	98.645	81.782	

Fig. 5.9 MSE evaluated in the training batch and test set at training time

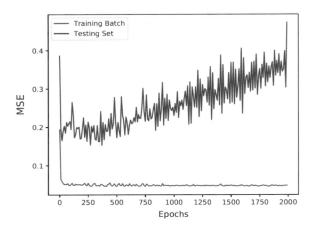

Deeper Networks with Polynomial Features

To optimize the training time, the depth of the network is iteratively increased. If an increase in performance from a shallower to a deeper network is not significative, the increase in training time might not be justifiable to have such a deep model. The same line of thought applies to the number of units in the hidden layers. In this regard, an ANN with a single hidden layer was trained to vary the number of hidden units and the results are shown in Table 5.5. These networks took from 10 to 30 min to train depending on the number of hidden neurons. The performance of the ANN with one hidden layer is considerably better than the simple linear regression (single-layer perceptron), hinting that the placement problem can be better modeled by a nonlinear function. The performance of these ANNs starts to decrease when the number of hidden units is over 500, pointing to the fact that there is no need to increase further the number of hidden units.

On the following test the depth of the network was changed to two hidden layers. These networks take 20–30 min to train, once again, depending on the number of hidden units, and their results are presented in Table 5.6. These networks, with two hidden layers, present better results than the ones with only one hidden layer, dropping the MAED error below 100 nm. Since the accuracies of the networks with two hidden layers are approximately 100%, it is necessary to lower the error and overlap thresholds that consider an output placement "correct", so that if the accuracy increases in further models it is possible to weight the differences between them. By decreasing these thresholds, the accuracy will naturally decrease. These were changed to 100 nm and 0 μm², respectively, meaning only placements with no overlap are considered "correct". The new accuracies for the two hidden layer networks are presented in Table 5.7.

Even though the thresholds for the accuracy measurment were reduced the accuracy metrics continue at high values. Meaning that the networks with two hidden layers (especially the one with 500 neurons in the first hidden layer and 250 in the second) have a high performance and output placement solutions extremely similar

Table 5.5 Performance of the ANN trained with 1 hidden layer

Hidden units	Test					Train				
	MAED (nm)	MOA (nm)	EA (%)	OA (%)	EOA (%)	MAED (nm)	MOA (nm)	EA (%)	OA (%)	EOA (%)
250	172.500	0.063	95.010	99.616	95.010	146.449	0.009	96.714	99.880	96.714
500	152.683	0.043	95.345	99.568	95.345	125.306	0.007	97.026	99.856	97.026
1000	165.801	0.057	95.345	99.664	95.345	145.348	0.007	97.098	99.892	97.098
1500	217.781	0.110	94.626	99.616	94.626	190.616	0.014	96.546	99.844	96.546

Table 5.6 Performance of ANNs trained with two hidden layers

Hidden units	Test					Train				
	MAED (nm)	MOA (nm)	EA (%)	OA (%)	EOA (%)	MAED (nm)	MOA (nm)	EA (%)	OA (%)	EOA (%)
500/100	74.815	0.021	98.656	99.856	98.656	58.793	0.015	99.244	99.940	99.244
500/250	57.588	0.031	98.896	99.856	98.896	37.464	0.015	99.868	99.916	99.868
550/470	90.703	0.058	98.273	99.760	98.273	72.336	0.009	99.232	99.928	99.232
1000/250	84.912	0.011	98.417	99.712	98.417	63.753	0.004	99.520	99.928	99.520
1000/500	85.417	0.010	98.560	99.760	98.560	66.886	0.002	99.556	99.844	99.556

Table 5.7 ANNs trained with two hidden layers (changed thresholds)

Hidden units	Test			Train		
	EA (%)	OA (%)	EOA (%)	EA (%)	OA (%)	EOA (%)
500/100	86.516	99.856	86.516	88.211	99.940	88.211
500/250	90.883	99.856	90.883	94.807	99.916	94.807
550/470	84.069	99.760	84.069	85.560	99.928	85.560
1000/250	86.372	99.712	86.372	89.614	99.928	86.614
1000/500	87.284	99.760	87.284	89.194	99.844	89.194

to the ones in the dataset. Furthermore, decreasing the overlap threshold to 0 did not have an impact on the accuracy. Since this threshold defines the border of acceptance of the placement prediction, the mean overlap area of the predicted placements that were considered being a match to their targets was actually 0. This means that in 99.896% of the examples on the dataset, the model outputs a placement where there is no overlap at all. To further reduce the MAED error, ANNs with three hidden layers were trained. The networks took 30–40 min to train, and the results of these networks are presented in Table 5.8. The increase in the depth of the network did not correspond to a direct increase in the performance of the model, which may indicate that the limit to what this type of networks can learn in this specific problem.

Networks with four hidden layers were trained in an attempt to reduce the error and raise the accuracy of the models. Since these networks are deeper, they took around 50–60 min to train. The results are shown in Table 5.9. The networks with four hidden layers brought a marginal increase in the performance of the model. Specifically, there were 2 different architectures that led to better results for different reasons. The network with 1200, 600, 300, and 100 hidden units in the respective hidden layers achieved a lower MAED error than the network with 1000, 500, 250, and 100 hidden units, but the latter achieved a higher EA and EOA. This means that the second network has a higher mean error, but outputs placements with lower error for more examples. Therefore, this is the network architecture that will be used for further testing with four hidden layers.

For further tests of the effect of additional preprocessing methods, the network architecture that had better performance of each different number of hidden layers case study is considered. To simplify the reference to each type of network, they are henceforward designated by the names shown in Table 5.10. The examples with the largest test error in each of the names ANNs are illustrated in Fig. 5.10. It is possible to observe that all correspond to the same sizing and that networks with 3 and 4 hidden layers are slightly closer to the target placement than the shallower networks.

Centering Data

To turn the patterns between the input and the output more evident, the dataset can be preprocessed before training the model so that the symmetry axis is located at $x = 0$. This will make the outputs that represent the x-coordinates of pair of devices symmetrical, which might make the patterns in the dataset easier for the

Table 5.8 Performance of the ANN trained with three hidden layers

Hidden units	Test						Train					
	MAED (nm)	MOA (nm)	EA (%)	OA (%)	EOA (%)		MAED (nm)	MOA (nm)	EA (%)	OA (%)	EOA (%)	
500/250/100	86.460	0.031	84.357	99.664	84.357		70.557	0.026	84.852	99.736	84.852	
500/250/50	85.051	0.048	83.541	99.856	83.541		70.427	0.002	85.872	99.940	85.872	
1000/500/250	70.700	0.070	88.532	99.760	88.532		52.889	0.019	90.849	99.856	90.849	
1250/600/300	87.530	0.059	88.292	99.568	88.292		67.030	0.003	92.180	99.952	92.180	

Table 5.9 Performance of the ANNs trained with four hidden layers

Hidden units	Test					Train				
	MAED (nm)	MOA (nm)	EA (%)	OA (%)	EOA (%)	MAED (nm)	MOA (nm)	EA (%)	OA (%)	EOA (%)
1200/600/300/100	53.363	0.029	92.658	99.904	92.658	33.476	0.017	96.714	99.940	96.714
1000/750/500/250	77.725	0.023	89.347	99.664	89.347	59.458	0.007	92.024	99.940	92.024
1000/500/250/100	67.629	0.039	94.194	99.904	94.194	48.351	0.003	98.045	99.928	98.045
1200/600/300/100	53.363	0.029	92.658	99.904	92.658	33.476	0.017	96.714	99.940	96.714

Table 5.10 Attributed names to each NN architecture

Name	Hidden layers	Test			
		Layer 1	Layer 2	Layer 3	Layer 4
ANN-1	1	500			
ANN-2	2	500	250		
ANN-3	3	1000	500	250	
ANN-4	4	1000	500	250	100

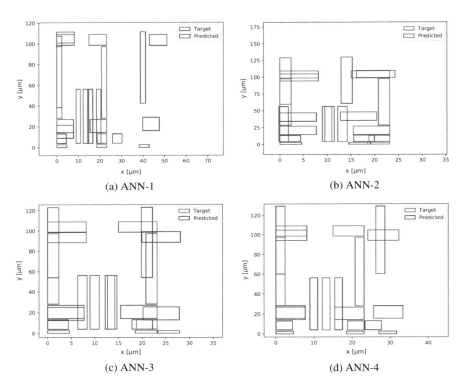

Fig. 5.10 Target and predicted placement for the examples with the largest error tested on every named ANN

network to learn. The results are shown in Table 5.11. Only ANN-1 and ANN-3 led to an improvement when the data were centered. ANN-3 with centering and ANN-4 without centering appear to be the networks that achieve a better performance, differing in MOA for about 1% in the test set. However, ANN-4 has a lower MAED. These are the only architectures that achieve a MOA superior to 90% and therefore selected for further tests. This is further highlighted in Fig. 5.11, where it is possible to observe that the placement with maximum error output by ANN-3 and ANN-4 is

Table 5.11 Performance of trained NNs with and without centering the data

ANN	Cent.	Test					Train				
		MAED (nm)	MOA (nm)	EA (%)	OA (%)	EOA (%)	MAED (nm)	MOA (nm)	EA (%)	OA (%)	EOA (%)
ANN-1	NO	167.502	0.029	46.689	99.472	46.689	137.886	0.009	49.376	99.868	49.376
	YES	145.843	0.088	52.351	99.472	52.351	121.516	0.011	55.313	99.844	55.313
ANN-2	NO	77.405	0.106	88.916	99.856	88.916	57.196	0.006	92.132	99.832	92.132
	YES	102.368	0.039	76.056	99.664	76.056	86.461	0.004	77.669	99.856	77.669
ANN-3	NO	93.513	0.041	85.077	99.568	85.077	77.502	0.014	87.395	99.844	87.395
	YES	65.513	0.044	94.146	99.808	94.146	45.868	0.0	98.369	100.0	98.369
ANN-4	NO	55.295	0.036	93.282	99.664	93.282	37.404	0.004	96.894	99.952	96.894
	YES	79.068	0.035	85.893	99.664	85.893	61.962	0.016	88.211	99.712	88.211

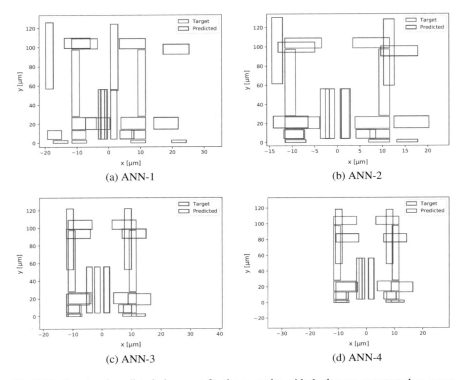

Fig. 5.11 Target and predicted placement for the examples with the largest error tested on every named ANN trained with centered data

close to the target. It is also possible to observe a minimal improvement in relation to the models trained without centered data.

5.3.3   Multiples Templates

To extrapolate for real-world application, the train and test sets were modified to take advantage of the different templates available on the dataset. In this first attempt, for each sizing, only the placement solution that corresponds to the placement with the smallest layout area was chosen among all the twelve templates. As some templates are better for certain devices' sizes ranges, the dataset will contain examples with multiple conflicting guidelines spread among training and test sets. This situation makes the patterns in the data harder to learn by the algorithm. To conduct these tests, the number of epochs was increased to 1500 since the optimization algorithm takes more time to converge. A higher number of conflicting guidelines was progressively added (2, 4, 8, and finally, 12), and the results are presented in Tables 5.12 and 5.13.

Table 5.12 Performance of the ANNs trained with and without centering the data, when the dataset has 2 or 4 different topological relations between cells

ANN	Cent.	Test										
		2 Templates						4 Templates				
		MAED (nm)	MOA (nm)	EA (%)	OA (%)	EOA (%)		MAED (nm)	MOA (nm)	EA (%)	OA (%)	EOA (%)
ANN-3	NO	249.934	0.169	75.528	97.553	75.528		558.915	3.761	6.622	92.850	6.622
	YES	234.179	0.220	73.704	97.553	73.704		429.202	3.574	65.211	93.138	65.211
ANN-4	NO	267.215	0.134	66.795	98.033	66.795		537.602	3.168	29.511	91.747	29.511
	YES	307.985	0.218	44.722	96.881	44.722		439.921	3.033	55.566	95.345	55.566

Table 5.13 Performance of the ANNs trained with and without centering the data, when the dataset has 8 or 12 different topological relations between cells

ANN	Cent.	Test					12 Templates				
		8 Templates									
		MAED (nm)	MOA (nm)	EA (%)	OA (%)	EOA (%)	MAED (nm)	MOA (nm)	EA (%)	OA (%)	EOA (%)
ANN-3	NO	1181.387	11.534	36.420	87.188	36.420	453.008	2.215	0.0	93.106	0.0
	YES	1115.585	12.013	0.048	79.223	0.048	206.533	0.476	21.913	97.393	21.913
ANN-4	NO	856.891	8.057	32.054	90.259	32.054	459.814	1.897	0.352	89.747	0.352
	YES	1179.797	11.926	0.0	80.326	0.0	497.691	3.359	2.079	85.333	2.079

Fig. 5.12 ANN-3 with
centered data and with 12
different templates: target
and predicted placement for
the examples with the largest
error

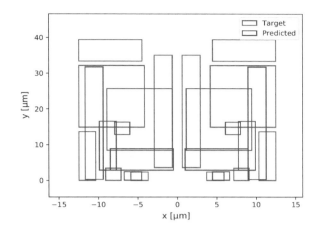

Since the existence of different conflicting guidelines in the dataset makes the patterns harder to learn, it is expected that the performance of the systems decreases significantly when comparing to the previous test cases where only one template was contemplated. ANN-3 with data centering achieves better performance in almost all tests. As the number of different templates increases, the performance of the model starts to decrease as expected, due to the variation in patterns it is trying to learn. Even though the output of the model is not optimal, it is still capable of outputting a meaningful placement as shown in Fig. 5.12. Although the overlap is high in this case, it is observable that it is trying to follow the patterns of a template, being almost completely symmetric. This fact means that the network learned the patterns of a template or subgroup of templates and applied to this example, even if not the intended originally by the test set.

5.3.4 Multiples Placement Output

To take a step further for making the ML system proposed in this work more suited for real-world application, the output of the model was changed to produce three different outputs: one with the smallest area, one with the maximum aspect ratio, and one with the minimum aspect ratio, as previously described. The distribution of each template in the different datasets used to train the network is presented in Table 5.14. Since there are some test cases where the maximum and/or minimum aspect ratio corresponds to almost one template, it is similar to train the network with a single template, and therefore, for that reason, it is expected a higher accuracy in those cases. The only networks that were trained were ANN-3 with centering and ANN-4 without centering, as they achieved better results when trained with datasets with more than one template. The results are shown in Tables 5.15 and 5.16.

Table 5.14 Distribution of each template in the datasets

Number of conflicting guidelines in the dataset

	2 Guidelines (ANN-2)			4 Guidelines (ANN-4)			8 Guidelines (ANN-8)			12 Guidelines (ANN-12)		
	Smaller Area (%)	Max. AR[a] (%)	Min. AR[a] (%)	Smaller area (%)	Max. AR[a] (%)	Min. AR[a] (%)	Smaller area (%)	Max. AR[a] (%)	Min. AR[a] (%)	Smaller area (%)	Max. AR[a] (%)	Min. AR[a] (%)
1	10.66	0.0	100.0	3.83	0.0	42.26	0.55	0.0	40.71	0.12	0.0	37.09
2	89.34	100.0	0.0	8.89	99.96	0.0	0.05	3.15	0.0	0.04	0.12	0.0
3				32.06	0.04	0.35	8.53	0.01	0.30	4.85	0.0	0.07
4				55.23	0.0	57.39	37.64	0.0	57.06	37.68	0.0	57.66
5							14.83	0.0	1.931	14.28	0.0	0.82
6							16.35	96.73	0.0	12.10	0.0	0.0
7							18.78	0.11	0.0	8.97	0.0	0.0
8							3.29	0.0	0.0	3.35	0.0	0.0
9										0.26	0.01	0.48
10										0.27	69.96	0.0
11										0.0	29.92	0.0
12										18.09	0.0	3.88

[a] AR refers to aspect ratio

Table 5.15 ANN-3 performance with centered data across the three placement output, with a different number of templates in the dataset

Templates	Test								
	Smaller			Maximum aspect ratio			Minimum aspect ratio		
	MAED (nm)	MOA (μm^2)	EOA (%)	MAED (nm)	MOA (μm^2)	EOA (%)	MAED (nm)	MOA (μm^2)	EOA (%)
1	58.103	0.068	90.355	57.981	0.056	90.499	58.058	0.056	90.307
2	238.624	0.172	81.574	90.495	0.044	87.716	105.242	0.053	82.006
4	467.690	4.500	57.821	91.599	0.071	85.797	404.523	4.145	47.073
8	893.245	10.538	16.891	248.068	1.784	79.271	434.566	5.330	52.783
12	297.888	1.256	11.548	300.019	0.443	6.878	206.100	0.528	31.078

Table 5.16 ANN-4 performance without centered data across the three placement output, with a different number of templates in the dataset

Templates	Test											
	Smaller			Maximum aspect ratio			Minimum aspect ratio		
	MAED (nm)	MOA (μm^2)	EOA (%)	MAED (nm)	MOA (μm^2)	EOA (%)	MAED (nm)	MOA (μm^2)	EOA (%)
1	61.193	0.032	91.603	61.199	0.032	91.603	61.188	0.032	91.603
2	235.945	0.142	83.301	110.155	0.068	86.756	96.260	0.055	85.413
4	457.402	3.583	48.417	130.160	0.052	81.574	411.164	3.526	41.315
8	887.761	8.640	26.008	322.172	1.551	67.658	454.046	3.118	46.113
12	562.028	3.849	0.016	603.367	1.949	0.304	404.998	1.179	0.240

ANN-3 outperforms ANN-4 is most of the test cases, especially when the number of templates in the dataset starts to increase.

These models have more difficulty learning patterns not only due to the high number of conflicting guidelines but also since the weights of the network are shared for the three proposed placements. However, the output provided is extremely promising, as they are producing valid or close-to-valid and meaningful placement solutions. Figure 5.13 illustrates this, as it corresponds to the median of the error of the ANN-3 with centered data and trained with 12 templates. In the case of ANN-4 without centering the data trained with 12 templates, the same situation is observed and presented in Fig. 5.14.

Table 5.17 shows the presence of each template in each dataset (training and testing) and the accuracy achieved in the examples of each template. As expected, the percentage of a certain template in the train set is directly correlated to the accuracy on the respective test examples of that template. The EOA values start to decrease when the number of templates used to train the network increases, as reported in the previous test cases with multiple templates but just one placement output. However, the higher values of error do not necessarily mean that the placement output by the

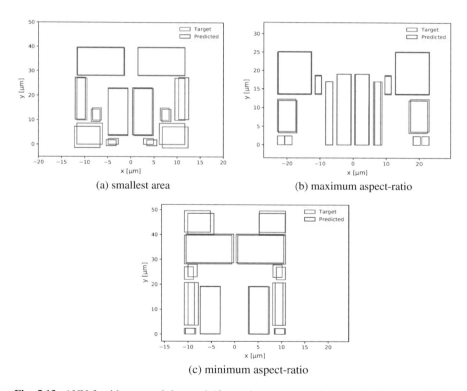

Fig. 5.13 ANN-3 with centered data and 12 templates: target and predicted placement for the example corresponding to the median error

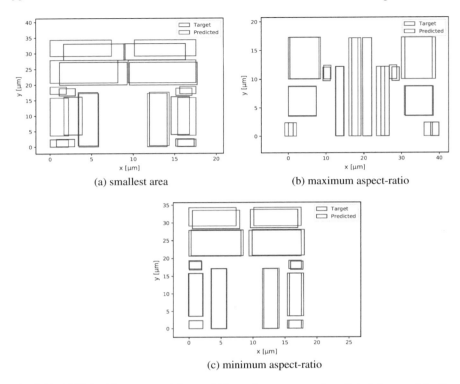

(a) smallest area   (b) maximum aspect-ratio

(c) minimum aspect-ratio

Fig. 5.14 ANN-4 without centered data and 12 templates: target and predicted placement for the example corresponding to the median error

network is not valid. As depicted in Fig. 5.15, the predicted placement, shown in (a), is very different from the target, shown in (b), and therefore has a high error associated but presenting a near-optimal symmetry and compaction. This difference in the output is caused by the fact that the network learned the patterns of one or more templates and then generalized the applied knowledge even though the intended placement solution from the test set is extremely different.

5.4   Conclusion and Future Research Directions

In this chapter, ML methodologies were used to develop ANNs that successfully predicted the layout placement of an amplifier, given their devices' sizes. This is a disruptive work, as no such approach has been taken in the field of analog IC placement, showing that a properly trained ANN can learn design patterns and generate placement solutions that are correct for sizing solutions outside the training data. This work only scratches the surface of the impact of ANNs and deep learning may

Table 5.17 Presence in each dataset and accuracy of each template

Temp.	Test								
	Smallest area			Maximum AR			Minimum AR		
	Presence in test dataset (%)	Presence in train dataset (%)	EOA (%)	Presence in test dataset (%)	Presence in train dataset (%)	EOA (%)	Presence in test dataset (%)	Presence in train dataset (%)	EOA (%)
1	0.160	0.116	0.0	0.0	0.0	–	37.876	37.091	21.284
2	0.0	0.036	–	0.064	0.116	0.0	0.0	0.0	–
3	5.326	4.845	0.0	0.0	0.0	–	0.016	0.068	0.0
4	37.716	37.683	21.416	0.0	0.0	–	56.734	57.660	40.569
5	14.571	14.276	1.207	0.0	0.0	–	0.912	0.816	0.0
6	12.332	12.101	1.167	0.0	0.0	–	0.0	0.0	–
7	8.541	8.971	0.562	0.0	0.0	–	0.0	0.0	–
8	2.767	3.350	0.0	0.0	0.0	–	0.0	0.0	–
9	0.336	0.264	0.0	0.016	0.008	0.0	0.416	0.484	0.0
10	0.208	0.272	0.0	70.250	69.957	6.444	0.0	0.0	–
11	0.0	0.0	–	29.671	29.919	7.925	0.0	0.0	–
12	18.042	18.086	17.199	0.0	0.0	–	4.047	3.882	0.0

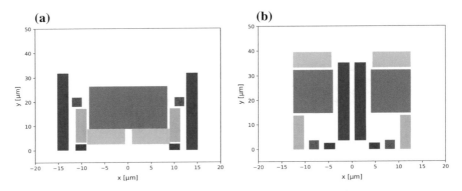

Fig. 5.15 ANN-3 with centering: **a** predicted and **b** target placement of the example with the largest error and smallest area

have in analog EDA, two major points are identified as future research directions: (1) although the MAED of a prediction provides an easy and valuable metric to assess if an ANN is capable of producing a floorplan for sizing solutions outside the training set, in this specific problem, large MAED values do not necessarily mean that the predicted placement is wrong, but instead a consequence of the ANN applying the acquired knowledge. Therefore, new metrics for the train and test of the ANN must be developed, and (2) the ultimate objective of this research is to lead to a model that could predict the placement of different circuit topologies (within certain classes of circuits and with a maximum number of devices). Thus, the architecture of the network must be changed, as each circuit can have a different number of devices. The way these circuit topologies are encoded in the input layer of the ANN is an ongoing research topic in analog EDA community.

References

1. H. Murata, K. Fujiyoshi, S. Nakatake, Kajitani. VLSI module placement based on rectangle-packing by the sequence-pair. IEEE Trans. Comput.-Aided Des. Integr. Circuits Syst. **15**(12), 1518–1524 (1996)
2. Y. Pang, F. Balasa, K. Lampaert, C.-K. Cheng, Block placement with symmetry constraints based on the o-tree non-slicing representation, in *Proceedings ACM/IEEE Design Automation Conference* (2000), pp. 464–467
3. Y.-C. Chang, Y.-W. Chang, G.-M. Wu, S.-W. Wu, B*-trees: A new representation for nonslicing floorplans, in *Proceedings of the 37th ACM/IEEE Design Automation Conference (DAC)* (2000), pp. 458–463
4. L. Jai-Ming, C. Yao-Wen, TCG: a transitive closure graph-based representation for non-slicing floorplans, in *Proceedings of the 38th ACM/IEEE Design Automation Conference (DAC)* (2001), pp. 764–769
5. P.-H. Lin, Y.-W. Chang, S.-C. Lin, Analog placement based on symmetry-island formulation. IEEE Trans. Comput. Aided Des. (TCAD) **28**(6), 791–804 (2009)

6. R. Martins, R. Póvoa, N. Lourenço, N. Horta, Current-flow and current-density-aware multi-objective optimization of analog IC placement. Integr. VLSI J. (2016). https://doi.org/10.1016/j.vlsi.2016.05.008
7. R. Martins, N. Lourenço, R. Póvoa N. Horta, On the exploration of design tradeoffs in analog IC placement with layout-dependent effects. in *International Conference on SMACD* (Lausanne, Switzerland, 2019)
8. Y. Yilmaz, G. Dundar, Analog layout generator for CMOS circuits. IEEE Trans. Comput.-Aided Des. Integr. Circuits Syst. (TCAD) **28**(1), 32–45 (2009)
9. R. Martins, N. Lourenco, N. Horta, Routing analog ICs using a multi-objective multi-constraint evolutionary approach. Analog. Integr. Circuits Signal Process. **78**(1), 123–135 (2013)
10. R. Martins, N. Lourenço, N. Horta, Multi-objective multi-constraint routing of analog ICs using a modified NSGA-II approach, in *International Conference on Synthesis, Modeling, Analysis and Simulation Methods and Applications to Circuit Design (SMACD)* (Seville, Spain, 2012), pp. 65–68
11. J. Cohn, J. Garrod, R.A. Rutenbar, L. Carley, Koan/Anagram II: new tools for device-level analog placement and routing. IEEE J. Solid-State Circ. (JSSC) **26**(3), 330–342 (1991)
12. K. Lampaert, G. Gielen, W. Sansen, A performance-driven placement tool for analog integrated circuits. IEEE J. Solid-State Circuits **30**(7), 773–780 (1995)
13. E. Malavasi, E. Charbon, E. Felt, A. Sangiovanni-Vincentelli, Automation of IC layout with analog constraints. IEEE Trans. Comput. Aided Des. Integr. Circ. Syst. (TCAD) **15**(8), 923–942 (1996)
14. N. Jangkrajarng, S. Bhattacharya, R. Hartono, C. Shi, IPRAIL—Intellectual property reuse-based analog IC layout automation. Integr. VLSI J. **36**(4), 237–262 (2003)
15. S. Bhattacharya, N. Jangkrajarng, C. Shi, Multilevel symmetry-constraint generation for retargeting large analog layouts. IEEE Trans. Comput. Aided Des. Integr. Circ. Syst. (TCAD) **25**(6), 945–960 (2006)
16. R. Martins, N. Lourenço, N. Horta, Laygen II—Automatic analog ICs layout generator based on a template approach, in *Genetic and Evolutionary Computation Conference (GECCO)* (Philadelphia, USA, 2012)
17. P.H. Wu, M.P.H. Lin, T.Y. Ho, Analog layout synthesis with knowledge mining, in *2015 European Conference on Circuit Theory and Design (ECCTD)* (2015), pp. 1–4
18. P.H. Wu, M.P.H. Lin, T.C. Chen, C.F. Yeh, X. Li, T.Y. Ho, A novel analog physical synthesis methodology integrating existent design expertise. IEEE Trans. Comput. Aided Des. Integr. Circuits Syst. **34**(2), 199–212 (2015)
19. R. Póvoa, et al., Single-stage amplifier biased by voltage-combiners with gain and energy-efficiency enhancement, in *IEEE Transactions on Circuits and Systems II: Express Briefs*, vol. 65, no. 3 (2018)
20. R. Martins, A. Canelas, N. Lourenço, N. Horta, On-the-fly exploration of placement templates for analog IC layout-aware sizing methodologies, in *2016 13th International Conference on Synthesis, Modeling, Analysis and Simulation Methods and Applications to Circuit Design (SMACD)*, (2016), pp. 1–4
21. R. Martins, N. Lourenço, A. Canelas, N. Horta, Stochastic-based placement template generator for analog IC layout-aware synthesis. Integr., VLSI J. **58**, 485–495 (2017)
22. D. Clevert, T. Unterthiner, S. Hochreiter, Fast and accurate deep network learning by exponential linear units (ELUs)," in *International Conference on Learning Representations* (2015), pp. 1–14
23. D. Kingma, J. Ba, Adam: a method for stochastic optimization, in *CoRR*, abs/1412.6980 (2014)
24. M. Abadi, et al., TensorFlow: large-scale machine learning on heterogeneous systems. 2015. Software available from tensorflow.org
25. Fabian Pedregosa et al., Scikit-learn: machine learning in python. J. Mach. Learn. Res. **12**, 2825–2830 (2011)
26. Python data analysis library. https://pandas.pydata.org/
27. Matplotlib. https://matplotlib.org/

Printed in the United States
By Bookmasters

Lecture Notes in Mechanical Engineering

Series Editors

Francisco Cavas-Martínez, Departamento de Estructuras, Universidad Politécnica de Cartagena, Cartagena, Murcia, Spain

Fakher Chaari, National School of Engineers, University of Sfax, Sfax, Tunisia

Francesco Gherardini, Dipartimento di Ingegneria, Università di Modena e Reggio Emilia, Modena, Italy

Mohamed Haddar, National School of Engineers of Sfax (ENIS), Sfax, Tunisia

Vitalii Ivanov, Department of Manufacturing Engineering Machine and Tools, Sumy State University, Sumy, Ukraine

Young W. Kwon, Department of Manufacturing Engineering and Aerospace Engineering, Graduate School of Engineering and Applied Science, Monterey, CA, USA

Justyna Trojanowska, Poznan University of Technology, Poznan, Poland

Lecture Notes in Mechanical Engineering (LNME) publishes the latest developments in Mechanical Engineering—quickly, informally and with high quality. Original research reported in proceedings and post-proceedings represents the core of LNME. Volumes published in LNME embrace all aspects, subfields and new challenges of mechanical engineering. Topics in the series include:

- Engineering Design
- Machinery and Machine Elements
- Mechanical Structures and Stress Analysis
- Automotive Engineering
- Engine Technology
- Aerospace Technology and Astronautics
- Nanotechnology and Microengineering
- Control, Robotics, Mechatronics
- MEMS
- Theoretical and Applied Mechanics
- Dynamical Systems, Control
- Fluid Mechanics
- Engineering Thermodynamics, Heat and Mass Transfer
- Manufacturing
- Precision Engineering, Instrumentation, Measurement
- Materials Engineering
- Tribology and Surface Technology

To submit a proposal or request further information, please contact the Springer Editor of your location:

China: Ms. Ella Zhang at ella.zhang@springer.com
India: Priya Vyas at priya.vyas@springer.com
Rest of Asia, Australia, New Zealand: Swati Meherishi
at swati.meherishi@springer.com
All other countries: Dr. Leontina Di Cecco at Leontina.dicecco@springer.com

To submit a proposal for a monograph, please check our Springer Tracts in Mechanical Engineering at http://www.springer.com/series/11693 or contact Leontina.dicecco@springer.com

Indexed by SCOPUS.

All books published in the series are submitted for consideration in Web of Science.

More information about this series at http://www.springer.com/series/11236

Carlo Vezzoli · Brenda Garcia Parra ·
Cindy Kohtala

Editors

Designing Sustainability for All

The Design of Sustainable Product-Service
Systems Applied to Distributed Economies

Springer

Editors
Carlo Vezzoli
Department of Design
Politecnico di Milano
Milan, Italy

Brenda Garcia Parra
Universidad Autónoma Metropolitana
Mexico City, Mexico

Cindy Kohtala
Department of Design
Aalto University School of Arts
Design and Architecture
Espoo, Finland

ISSN 2195-4356   ISSN 2195-4364   (electronic)
Lecture Notes in Mechanical Engineering
ISBN 978-3-030-66299-8   ISBN 978-3-030-66300-1   (eBook)
https://doi.org/10.1007/978-3-030-66300-1

This Springer imprint is published by the registered company Springer Nature Switzerland AG
The registered company address is: Gewerbestrasse 11, 6330 Cham, Switzerland

Foreword

During the last few decades, the history of design culture and practice, when dealing with the issue of sustainability, has moved from individual products to systems of consumption and production, and from strictly environmental problems to a complex blend of socio-ethical, environmental and economic issues. In this context, a clear challenge is to provide Sustainability for All (accessible even within low- and middle-income contexts), coupling environmental protection with social equity, social cohesion and economic prosperity. Within this framework, it is crucial that design can take a proactive role and become an agent to extend access to sustainable solutions. Design can do so because within its genetic code, by principle its role is to improve the quality of the world: an ethical–cultural component that, though not generally apparent, can be found in a deeper examination of the majority of designers' motivations. Finally, it is evident that a key role has to be played by Higher Education Institutions, both in researching and defining the new roles designers may play, as well as in curricular proposals where a new generation of design should grow. A challenging journey is ahead of us. And from this perspective, we believe this book will contribute to a larger change in the design community invited to meet this challenge.

Milan, Italy   Carlo Vezzoli
Mexico City, Mexico   Brenda Garcia Parra
Helsinki, Finland   Cindy Kohtala
London, UK   Fabrizio Ceschin
Curitiba, Brazil   Aguinaldo dos Santos
Recife, Brazil   Leonardo Castillo
Bangalore, India   Ranjani Balasubramanian
Guwahati, India   Ravi Mokashi Punekar
Beijing, China   Xin Liu
Changsha, China   Jun Zhang
Cape Town, South Africa   Ephias Ruhode
Cape Town, South Africa   Elmarie Costandius
Delft, Netherlands   JC Diehl

Acknowledgements

This volume is a collaboration of the following authors representing all partners in the LeNSin project.

Carlo Vezzoli[1] wrote Sects. 1.1, 1.2, 1.4, 1.5, 2.3, 2.4.6, 2.4.7, 2.5, 2.7, 3.1–3.3, 4.1, 4.2, 4.3.1–4.3.5.

Brenda Garcia Parra[2] and Sandra Molina Mata (see Footnote 2) wrote Sects. 2.8.5, 5.6, 5.7.

Cindy Kohtala[3] wrote Sects. 2.2, 2.7, 2.8.3.

Tatu Marttila (see Footnote 3) and Cindy Kohtala wrote Sects. 3.2.5, 5.1, 5.2, 5.7.

Fabrizio Ceschin[4] wrote Sects. 1.3, 2.4.3, 3.2.1.

Aine Petrulaityte (see Footnote 4) wrote Sects. 2.4.2, 4.3.7.

Aguinaldo dos Santos[5] wrote Sect. 5.3.

Aguinaldo dos Santos and Gabriela Garcez Duarte (see Footnote 5) wrote Sects. 2.1, 2.6, 2.7, 2.8.1.

Aguinaldo dos Santos and Isadora Burmeister Dickie[6] wrote Sect. 2.4.1.

Ranjani Balasubramanian[7] wrote Sects. 2.8.4, 3.1, 3.2.1–3.2.3, 3.2.5, 3.2.7, 3.4.

Ranjani Balasubramanian, Jacob Matthew[8], Abhijit Sinha[9] and Christoph Neusiedl (see Footnote 9) wrote Sects. 3.2.4, 3.2.6.

Fang Zhong[10], Nan Xia (see Footnote 10) and Xin Liu (see Footnote 10) and Jun Zhang[11] wrote Sect. 5.5.

---

[1] Politecnico di Milano, Design Department, Italy.

[2] Universidad Autónoma Metropolitana, Mexico.

[3] Aalto University School of Arts, Design and Architecture, Department of Design, Finland.

[4] Brunel University London, Department of Design, UK.

[5] Universidade Federal do Paraná, Brazil.

[6] Universidade da Região de Joinville (Univille), Brazil.

[7] Srishti Institute of Art, Design and Technology, India.

[8] Industree Foundation, India.

[9] Project DEFY, India.

[10] Tsinghua University, China.

[11] Hunan University, China.

Nan Xia wrote Sects. 2.4.4, 2.8.2.

Sharmistha Banerjee, Pankaj Upadhyay and Ravi Mokashi Punekar[12] wrote Sects. 4.3.6, 5.4.

Leonardo Castillo[13] and Carla Pasa Gómez wrote Sects. 2.4.5, 5.3.

JC Diehl[14] contributed to Sects. 1.1, 1.2, 2.1, 2.5–2.7, 4.2, 5.3.

Ephias Ruhode[15] and Corbin Raymond[16] contributed to Sects. 3.1, 5.3.

Luca Macrì[17] wrote case studies in Sects. 3.2.1–3.2.5, 3.2.7, 3.3.

[12]Indian Institute of Technology Guwahati, India.

[13]Universidade Federal de Pernambuco, Brazil.

[14]Delft University of Technology, the Netherlands.

[15]Cape Peninsula University of Technology, South Africa.

[16]Stellenbosch University, South Africa.

[17]Politecnico di Milano, Design Department, Italy.

Introduction

One major issue attached to the transition towards a sustainable society is that of improving **social equity and cohesion**, while empowering **locally based** enterprises and initiatives, for an **environmentally** sustainable re-globalization process characterized by a **democratization of access to resources, goods** and **services**. For a just and sustainable world, this necessitates attention to capacitating low- and middle-income contexts, regions or social groups and enabling access to resources, at the same time as enhancing local capabilities in high income regions in a way that does not exploit the poor and vulnerable. Two promising and interwoven offer models coupling environmental with economic and social sustainability are Sustainable Product-Service Systems (S.PSS) and Distributed Economies (DE). In relation to these two models, a new promising Research Hypothesis has been proposed, studied and characterized during the LeNSin Erasmus+ European Union funded project[18] and a new promising role of and for design has been envisioned. The outcomes of this research endeavour are elaborated in this book.

Firstly, the concept of Sustainable Product-Service Systems (S.PSS) is introduced as a known win-win offer model for sustainability. The idea of Distributed Economies (DE) is then introduced as a promising offer model for locally based sustainability. This is followed by an elaboration of the following scenario: S.PSS applied to DE as an opportune approach to diffusing sustainability for all. A scenario of S.PSS applied to DE is presented together with illustrative case studies. Finally, a new role for design in developing the S.PSS applied to DE model is presented, i.e. System Design for Sustainability for All (SD4SA).

---

[18]LeNSin, the International Learning Networks on Sustainability is an EU-supported (ERASMUS+) project. It aims to promote a new generation of designers and educators capable of effectively contributing to the transition towards a sustainable society for all. The partnership includes 36 universities and institutions (14 partners and all other associated partners) from Brazil, China, India, Mexico, South Africa and, in Europe, Finland, Italy, The Netherlands and the United Kingdom. It is part of the LeNS network established in 2007, now involving more than 150 Higher Education Institutions from all continents, that adopt a learning-by-sharing approach to knowledge generation and dissemination, with an open access ethos. www.lens-international.org.

Contents

Product-Service Systems Development for Sustainability. A New Understanding

Carlo Vezzoli, Fabrizio Ceschin, and Jan Carel Diehl

1 The Role of PSS in Addressing Sustainability

1.1 The Sustainability Challenge

In 1972 the book *Limits to Growth* was published based on a first computerized simulation of the effects on nature of the ongoing system of production and consumption [19]. It was the first scientific forecast of a possible global eco-system collapse. Fifteen years later, in 1987, the United Nations (UN) World Commission for Environment and Development (WCED) provided the first definition of Sustainable Development: A social, and productive development that takes place within the limits set by "nature" and meets the needs of the present without compromising those of the future generation within a worldwide equitable redistribution of resources. This also incorporates the fundamental challenge of social equity and cohesion (i.e. the socio-ethical dimension of sustainability). In the autumn of 2015, the UN updated the commitments, goals and actions for sustainable development by approving the "Agenda 2030 for Sustainable Development" [28] as a mutual commitment to global development, in favour of human well-being and to preserve the environment. The main outputs of the Agenda are the 17 Sustainable Development Goals (SDGs), which gather together the main challenges to be achieved by 2030 in relation to the three dimensions of sustainable development, i.e. the environmental protection, the social inclusion and the economic prosperity.

C. Vezzoli
Design Department, Politecnico di Milano, Milan, Italy

F. Ceschin
Department of Design, Brunel University London, London, UK

J. C. Diehl (✉)
Design for Sustainability, Delft University of Technology, Delft, The Netherlands
e-mail: j.c.diehl@tudelft.nl

© The Author(s) 2021
C. Vezzoli et al. (eds.), *Designing Sustainability for All*, Lecture Notes
in Mechanical Engineering, https://doi.org/10.1007/978-3-030-66300-1_1

1

It is within this framework that this book presents Sustainable Product-Service System (S.PSS) and Distributed Economies (DE) as key promising and interwoven offer models coupling environmental and social with economic sustainability. Moreover, S.PSS applied to DE is a promising approach to diffuse sustainability in low- and middle-income contexts. This volume also elaborates on the role design can play to generate new ideas and solutions addressing S.PSS applied to DE, as well as develop and diffuse related solutions, i.e. designing sustainability for all. This chapter presents an updated understanding of how PSS addresses Sustainability and the role of design.

1.2   Sustainable Product-Service System: A Win-Win Opportunity for Sustainability

A key contemporary query is the following: within the entangled and complex environmental, social and economic crises, where are the opportunities? Do we know any offer or business model capable of creating (new) value, decoupling it from material and energy consumption? In other words, significantly reducing the environmental impact of traditional production/consumption systems? In fact, the concept of Sustainable Product-Service System (S.PSS) has been studied since the end of the 1990s [10, 12, 18, 20, 23, 29] as a promising offer/business model in this regard. More recently, S.PSS has been demonstrated [32, 36] to be a clearly promising offer model to extend the access to good and services even to low- and middle-income contexts, thus enhancing social equity and cohesion as well. Finally, it is a win-win offer model combining the three dimensions of sustainability, the economic with the environmental and the socio-ethical. An S.PSS can be defined as follows [36]:

> Sustainable Product-Service System (S.PSS) is an offer model providing an integrated mix of products and services that are together able to fulfil a particular customer/user demand (to deliver a "unit of satisfaction"), based on innovative interactions between the stakeholders of the value production system (satisfaction system), where the ownership of the product/s and/or the life cycle services costs/responsibilities remain with the provider/s, so that the same provider/s continuously seek/s environmentally and/or socio-ethically beneficial new solutions, with economic benefits.

S.PSSs are value propositions introducing considerable innovation on different levels (see also Fig. 1):

- They shift the business focus from selling (only) **products** to offering a so-called **"unit of satisfaction"**,[1] i.e. a combination of products and services jointly capable of achieving an ultimate user satisfaction.
- They shift the value perceived by the customer/end-user from **individual ownership** to **access** to goods and services.
- They shift the primary innovation from a **technological** one to an innovation on a **stakeholder interaction** level.

This approach is also supported by the European Union in its action plan on the Circular Economy, when stating that: "incentivising product-as-a-service or other models where producers keep the ownership of the product or the responsibility for its performance throughout its lifecycle" [10]. Finally, in the key understanding of our discourse, S.PSSs are offer models with a win-win sustainability potential, i.e. they are offer/business models capable of creating (new) value, decoupling it from resource consumption and increase of negative environmental impact whilst extending access to good and services to low- and middle-income people and, at the same time, enhancing social equity and cohesion.

1.3 PSS Types

Three main S.PSS approaches to system innovation have been studied and listed as favourable for eco-efficiency [13, 26, 29, 37]:

1. *Product-oriented S.PSS*: services providing added value to the product life cycle.
2. *Use-oriented S.PSS*: services providing "enabling platforms" for customers.
3. *Result-oriented S.PSS*: services providing "final results" for customers.

Product-oriented S.PSS: adding value to the product life cycle (type I)

Let us start with an example of an eco-efficient system innovation adding value to the product life cycle.

TRADITIONAL PRODUCT SALE		TO S.PSS	
SELLING	PRODUCT	TO "UNIT OF SATISFACTION"	
INNOVATION	TECHNOLOGICAL	TO STAKEHOLDER CONFIGURATION	S.PSS
CUSTOM. VALUE	INDIVIDUAL OWNERSHIP	TO ACCESS	

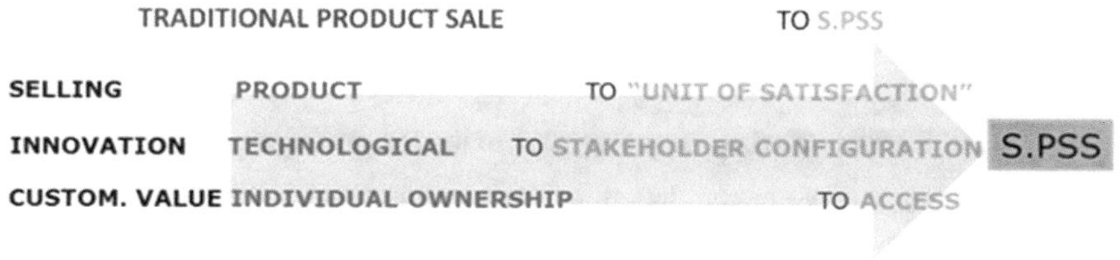

Fig. 1 S.PSS: a paradigm shift from a traditional product offer

---

[1]A satisfaction unit can be defined as [36] "a defined (quantified) satisfaction of a customer that could be fulfilled by one or more mix of products and services, used as a reference unit to design and to evaluate the sustainability benefits and impacts".

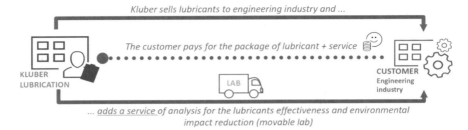

Fig. 2 Klüber lubricants service. Main company-customer interactions

Klüber lubricants service
Klüber offers lubricants plus the service for on-site identification of equipment inefficiency, and the potential reduction of emissions' impact (see Fig. 2). This innovative interaction between the company and the customer adding all-inclusive life cycle services allows the company's economic interest to be different from only selling more lubricants.

In summary, a ***Product-oriented S.PSS*** *innovation* adding value to the product life cycle is defined as:

a company/organization (alliance of companies/organizations) that provides all-inclusive life cycle services – maintenance, repair, upgrading, substitution and product take-back – to guarantee the life cycle performance of the product/semi-finished product (sold to the customer/user).

A typical service contract would include maintenance, repair, upgrading, substitution and product take-back services over a specified period of time. The customer/user responsibility is reduced to the use and/or disposal of the product/semi-finished product (owned by the customer), since she/he pays all-inclusively for the product with its life cycle services, and the innovative interaction between the company/organization and the customer/user drives the company/organization's economic interest in continuously seeking environmentally beneficial new solutions, i.e. the economic interest becomes something other than only selling a larger amount of products.

Use-oriented S.PSS: offering enabling platforms for customers (type II)

The following box describes an example of an eco-efficient system innovation as enabling platforms for customers.

Riversimple. Pay-per-month mobility solution
Riversimple provides a pay-per-month ownerless car with all-inclusive energy, maintenance, repair, insurance and end-of-life collection (see Fig. 3). This

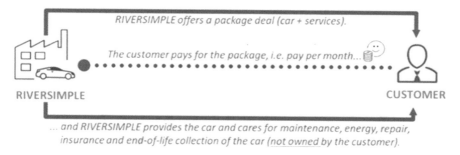

Fig. 3 Riversimple. Pay-per-month mobility solution

type of innovative interaction between the company and the customer (owner-less car with all-inclusive life cycle services) promotes the provider's economic interest to foster the design or offer of a long-lasting, energy-efficient and recyclable car.

In summary, a ***use-oriented S.PSS*** *innovation* offering an enabling platform to customers is defined as:

a company/organization (alliance of companies/organizations) that provides access to products, tools and opportunities enabling the customer to get their "satisfaction". The customer/user does not own the product/s but operates them to obtain a specific "satisfaction" (and pays only for the use of the product/s).

Depending on the contract agreement, the customer/user could have the right to hold the product/s for a given period of time (several continuous uses) or only for one use.

Commercial structures for providing such services include leasing, pooling or sharing of certain goods for a specific use. The customer/user consequently does not own the products, but operates on them to obtain a specific final satisfaction (the client pays for the use of the product). Again, in this case, the innovative interaction between the company/organization and the customer/user drives the company/organization to continuously seek environmentally beneficial new solutions together with economic benefits, e.g. to design highly efficient, long-lasting, reusable and recyclable products.

Result-oriented S.PSS: offering final results to customers (type III)

The following describes an example of an eco-efficient system innovation providing final results to customers.

Philips. Pay-per-Lux
The customer pays to have an agreed amount of lighting (lux) in its building. Philips is responsible for (paying for) the installation, upgrading, repair and

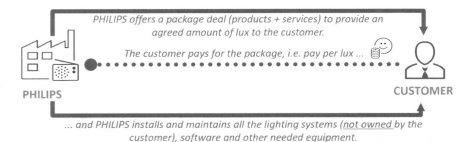

Fig. 4 Philips. Pay-per-Lux

end-of-life collection of all products/equipment (owned by Philips) (Fig. 4). This kind of innovative interaction between the company and the customer (an ownerless lighting system with all-inclusive life cycle services) encourages the provider's economic interest to foster the design/offer of long-lasting, reusable and recyclable lighting systems.

A *result-oriented S.PSS innovation* offering final results to customers can be defined as:

a company/organization (alliance of companies/organizations) that offers a customized mix of services, instead of products, in order to provide a specific final result to the customer. The customer/user does not own the products and does not operate on them to obtain the final satisfaction (the customer pays the company/organization to provide the agreed results).

The customer/user benefits by being freed from the problems and costs involved in the acquisition, use and maintenance of equipment and products. The innovative interaction between the company and the customer/user drives the company's economic and competitive interest to continuously seek environmentally beneficial new solutions, e.g. long-lasting, reusable and recyclable products. Moreover, if properly conceived, S.PSS can offer to low- and middle-income people the possibility to have access to services that traditional product sales models would not allow (i.e. by lower initial costs).

In fact, it has been argued that in low- and middle-income contexts "an S.PSS innovation may act as a business opportunity to facilitate the process of socio-economic development by jumping over the stage characterized by individual consumption/ownership of mass-produced goods towards a 'satisfaction-based' and 'low resource-intensity' advanced service-economy" [29].

1.4 S.PSS Environmental Benefits

When is an S.PSS eco-efficient? Better still, when is an S.PSS decoupling the economic interests from both an increase in resource consumption and a decrease of demaging environmental impacts?

In other words, why and when is an S.PSS producer/provider economically interested in design for environmental sustainability? The following S.PSS environmental and economic win-win benefits could be highlighted (adapted from [36]):

(a) *Product lifetime extension*: As far as the S.PSS provider is offering the products retaining the ownership and being paid per unit of satisfaction, or offering all-inclusive the product with its maintenance, repair, upgrade and substitution, the **longer** the product/s or its components last (environmental benefits), and the **more** the producer/provider avoids or postpones the disposal costs plus the costs of pre-production, production and distribution[2] of a new product substituting the one disposed of (economic benefits). Hence the producer/providers are driven by economic interests to design (offer) for lifespan extension of product/s (with eco-efficient product Life Cycle Design (LCD) implications) (Fig. 5).

(b) *Intensive use of product*: As far as the S.PSS provider is selling a shared use of products (or product's components) to various users, the **more** intensively the product/s (or some product's components) are used, i.e. the more time (environmental benefits), the **higher** the profit, i.e. proportionally to the overall use time (economic benefits). Hence, the producer/providers are driven by economic interests to design for intensive use of product/s (eco-efficient product LCD implications) (Fig. 6).

(c) *Resource consumption minimization*: As far as the S.PSS provider is selling all-inclusive the access to products and the resources it consumes in use, with payment based on unit of satisfaction (product's ownership by the producer/provider), the **higher** the product/s resource efficiency in use is (environmental benefits), and the **higher** the profit, i.e. the payment minus (among others) the costs of resources in use (economic benefits). Hence, the producer/provider is driven by economic interests to design/offer product/s minimizing resource consumption in use (eco-efficient product LCD implications) (Fig. 7).[3]

(d) *Resources' renewability*: When the S.PSS provider has an all-inclusive offer of a utility, with pay per period/time/satisfaction (e.g. energy production unit ownership by the producer/supplier), the **higher** the proportion of passive/renewable sources is in relation to non-passive/non-renewable (environmental benefits), and the **higher** the profit, i.e. the payment minus (among others) the costs of non-passive/non-renewable sources (economic benefits). Hence, the producer/provider is driven by economic interests to design (offer) for

---

[2]Even marketing and advertisement costs could be avoided.
[3]Resource efficiency might include the end-of-life stage (recycling, re-use, composting, etc.) where it would be of interest to the S.PSS provider to make this stage also economically relevant.

passive/renewable resource optimization (eco-efficient product LCD implications) (Fig. 8).

(e) *Material life extension*: As far as the S.PSS provider is selling the product all-inclusive with its end-of-life treatment/s, the **more** the materials are either recycled, incinerated with energy recovery or composted (environmental benefits), the **more** costs are avoided of both landfilling and either the purchase of new primary material, energy or compost (economic benefits). Hence, the producer/provider is driven by economic interests to design for material life extension, i.e. recycling, energy recovery or composting (eco-efficient product LCD implications) (Fig. 9).

(f) *Minimization of toxicity and harmfulness*: As far as the S.PSS provider is selling toxic or harmful product/s all-inclusive with use and/or end-of-life toxicity/harm management services, the **lower** the potential toxic or harmful emissions are in use and/or at the end-of-life (environmental benefits), the **more** costs are avoided of both toxic/harmful treatments in use and/or at the end-of-life. Hence, the producer/provider is driven by economic interests to design (offer) for toxicity/harm minimization (eco-efficient product LCD implications) (Fig. 10).

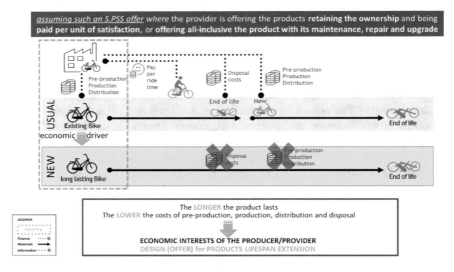

Fig. 5 S.PSS model fostering the design (offer) for product lifespan extension

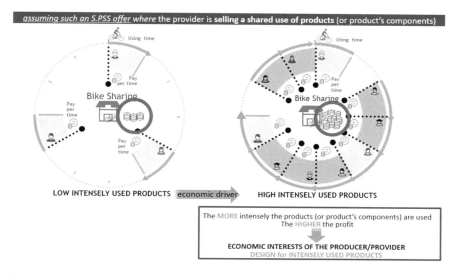

Fig. 6 S.PSS model fostering the design (offer) for intensive use of the product

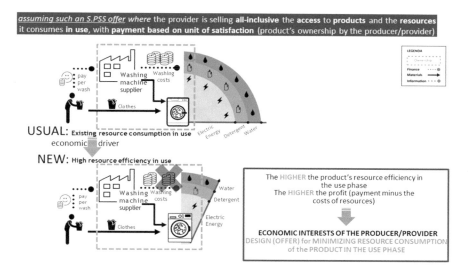

Fig. 7 S.PSS model fostering the design (offer) of products minimizing resource consumption in the use phase

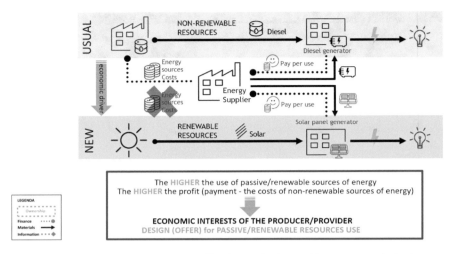

Fig. 8 S.PSS model fostering the design (offer) for passive/renewable resource optimization

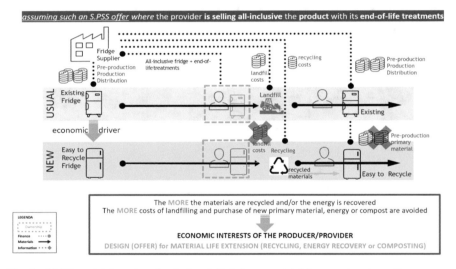

Fig. 9 S.PSS model fostering the design (offer) for material life extension (recycling, energy recovery or composting)

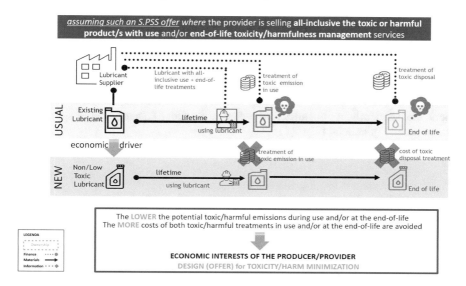

Fig. 10 S.PSS model fostering the design (offer) for toxicity/harm minimization

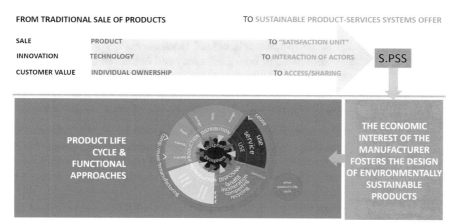

Fig. 11 S.PSS as a model making product Life Cycle Design economically relevant for the manufacturer/provider

To conclude, when is an S.PSS eco-efficient? When the product ownership and/or the economic responsibility for its life cycle performance remains with the producers/providers who are selling a unit of satisfaction rather than (only) the product. And why does this happen? Because this way, we shift or allocate the direct economic and competitive interest to reduce the products' and/or the services' environmental impacts, onto the stakeholder responsible for their design and development. Consequently, within an S.PSS model, a product LCD/eco-design approach is economically beneficial (Fig. 11).

In other words, an S.PSS producer/provider is economically interested in design for:

- product lifespan extension and use intensification;
- material life extension (recycling, energy recovery, composting);
- material consumption minimizations;
- energy consumption minimizations;
- resources' (materials and energy) renewability/biocompatibility;
- resources' (materials and energy) toxicity/harmfulness minimization.

1.5   S.PSS Socio-Ethical Benefits

Why may S.PSS foster socio-ethical benefits? Because S.PSS make goods and services economically accessible to both final users and entrepreneurs, also in low- and middle-income contexts. The following S.PSS socio-ethical and economic win-win benefits could be highlighted (updated from [36]): The first two are related to end-users and the third, fourth and fifth are related to entrepreneurs/organizations.

(a) *End-user product accessibility*: As far as the S.PSS model is selling the access rather than mere product ownership, this reduces or avoids purchasing costs of products that are frequently too high for low- and middle-income end-users *(economic benefits),* i.e. making goods and services more easily accessible (*socio-ethical benefits*) (Fig. 12).

(b) *Reduction of interrupted product use*: As far as the S.PSS model is selling the 'unit of satisfaction' including life cycle services costs, this reduces or avoids running costs for maintenance, repair, upgrade, etc. that are too high for low- and middle-income end-users (economic benefits), i.e. who can avoid interruption of product use (socio-ethical benefits) (Fig. 13).

(c) *Enterpreneurs/organizations' equipment accessibility*: As far as the S.PSS model is selling access rather than the (working) equipment itself, this reduces or avoids initial (capital) investment costs of equipment, which are frequently too high for low- and middle-income entrepreneurs/organizations (economic benefits), i.e. facilitating new business start-ups in low- and middle-income contexts (socio-ethical benefits) (Fig. 14).

(d) *Reduction of interrupted equipment use*: As far as the S.PSS model is selling all-inclusive life cycle services with the equipment offer to entrepreneurs, this reduces or avoids running costs for equipment maintenance, repair, upgrade, etc. that are frequently too high for low- and middle-income entrepreneurs/organizations (economic benefits), i.e. this avoids interruption of equipment use and subsequently working activities (socio-ethical benefits) (Fig. 15).

(e) *Local employment and competencies improvement*: As far as the S.PSS model is offering goods and services without product purchasing costs, they open new

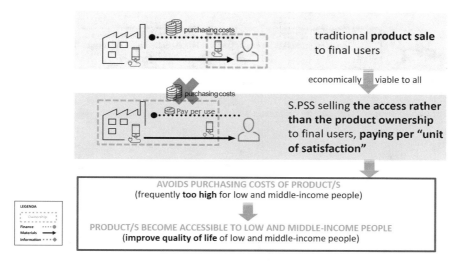

Fig. 12 S.PSS model making product/s accessible to low- and middle-income end-users

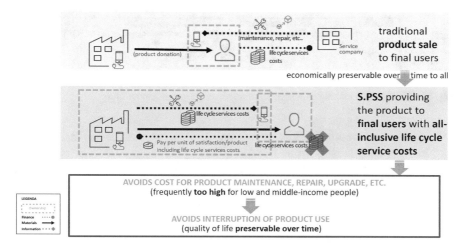

Fig. 13 S.PSS model making quality of life preservable over time in low- and middle-income contexts

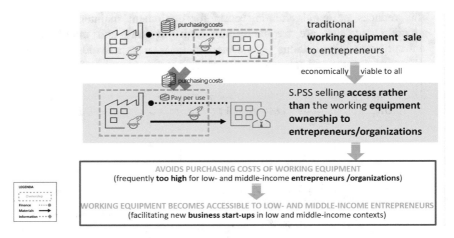

Fig. 14 S.PSS model facilitating new business start-ups in low- and middle-income contexts

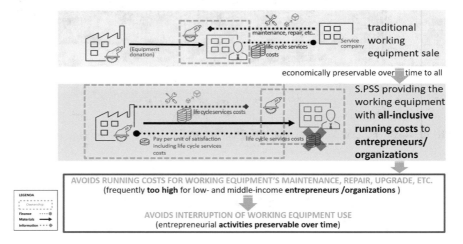

Fig. 15 S.PSS model making entrepreneurial activities preservable over time

market opportunities for local entrepreneurs via new potential low- and middle-income customers (such as Bottom of the Pyramid, or BoP), i.e. potentially empowering locally based economies and life quality (socio-ethical benefits) (Fig. 16).

The service dimension of an S.PSS demands local providers, thus generating local jobs. This contributes directly to social cohesion, as it reduces the need for migration or long commutes; increases the likelihood of better balance between work and social life; and thus provides a context where the social fabric can be built up and/or consolidated.

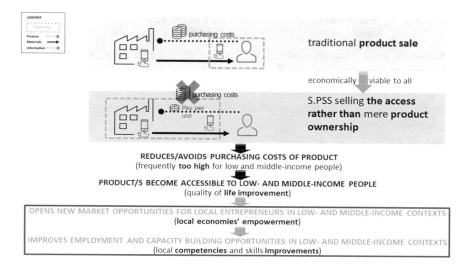

Fig. 16 S.PSS model improving local employment, competencies and skills

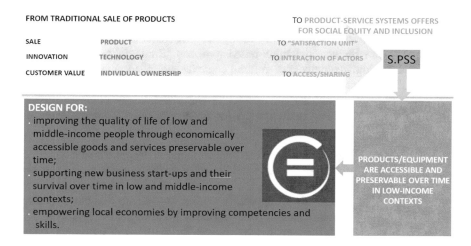

Fig. 17 S.PSS model improving local life quality, competencies and skills

Finally, within an S.PSS model the producer/provider is economically interested in design for social equity, i.e. to extend sustainable access to products/equipment for low- and middle-income people (see Fig. 17), by designing for:

- improving the quality of life of low- and middle-income people through economically accessible goods and services preservable over time;
- supporting new business start-ups and their survival over time in low- and middle-income contexts;
- empowering local economies by improving competencies and skills.

1.6   S.PSS Economic and Competitive Benefits

What are the main economic and competitive benefits of S.PSS? The following S.PSS economic and competitive benefits could be highlighted [36]:

- As far as the S.PSS model offers service along all its life cycle, organizations can establish longer and stronger relationships with customers, i.e. increasing customer loyalty;
- As far as the S.PSS models are different offers from traditional product sales, which are nowadays in saturated markets, they can open up new business opportunities, i.e. empowering strategic positioning;
- As far as the S.PSS model offers goods and services without initial investment costs, they open new market opportunities for middle- and low-income people (BoP), i.e. empowering locally based economies.

2   PSS Design for Sustainability

The introduction of PSS innovation for sustainability into design has led design researchers to work on defining new skills of a more strategic nature [2, 3, 18, 25, 37], which aim at system sustainability through a convergence of stakeholder interests and are coherent with the satisfaction-based approach. 'Strategic' here also refers to the necessary acknowledgement of cultural contexts and inherent opportunities and barriers built into the social fabric.

In relation to the characteristics of S.PSS described in the previous section, three main approaches and related skills for Product-Service System Design for Sustainability could be highlighted [36]:

- a **"satisfaction-system"** approach: calling for skills to design the satisfaction of a particular demand (a "satisfaction unit") and hence all its related products and services;
- a **"stakeholder configuration"** approach: calling for skills to design the interactions of the stakeholders of a particular satisfaction-system;
- a **"system sustainability"** approach: calling for skills to design such stakeholder interactions (offer model) that make the providers economically interested to continuously seek both environmentally and socio-ethical new beneficial solutions.

The first key point lies in the satisfaction-based approach, where the focus is no longer on delivering a single product. It is thus inadequate to merely design or assess a single product, but instead we consider the whole process of every product and service associated with satisfying certain needs and/or desires. The second key task is to introduce a stakeholder configuration approach. If we want to design the stakeholder interactions, the system design approach should project and promote innovative types of interactions and partnerships between appropriate socio-economic stakeholders, while responding to a particular social demand for satisfaction. Therefore, designing the configuration of a system means understanding what stakeholder profiles should be in place and what the best interrelationships are, in the sense of financial, resource, information or labour flows. Last but not least, it must be emphasized that, as stated by various authors [3, 20, 29–31], not all PSS innovations are driven by the economic interest to have a reduced environmental impact, nor do they necessarily promote social equity and cohesion. For this reason, it is expedient to operate and adopt appropriate criteria and guidelines in the design process towards sustainable stakeholder interactions/relationships. Having understood this, **Product-Service System design for sustainability** can be defined as follows (adapted from [36]):

the design of the system of products and services that are together able to fulfil a particular customer demand (to deliver a "unit of satisfaction"), based on the design of innovative interactions between the stakeholders of the value production system (satisfaction system), where the ownership of the product/s and/or the life cycle services costs/responsibilities remain with the provider/s, so that the same provider/s continuously seek/s environmentally and/or socio-ethically beneficial new solutions, together with economic benefits.

3 S.PSS in Relation to Other Design for Sustainability Approaches

This book focuses on S.PSS and the role it can play in fostering Distributed Economies. However, it remains essential to discuss the linkages between S.PSS and other Design-for-Sustainability (DfS) approaches. In fact, in order to exploit the sustainability potential of PSS solutions, other DfS approaches should be adopted and used in combination with S.PSS design [7].

To begin with, it is important to highlight that the sustainability profile of a PSS strictly depends on how the products included in the offer have been designed. It is true that through an S.PSS approach it is possible to develop business models in which the manufacturer and the other stakeholders involved in the solution have a potential economic incentive to take responsibility for the PSS life cycle and optimize material and energy consumption (see Sect. 1.4). However, in order to exploit this sustainability potential, the products need to be correctly designed. For example, in a use-oriented PSS (see Sect. 1.3), in which manufacturers keep ownership of products and offer access to them, products need to be designed to be long-lasting (considering also the shared use), easy to maintain and repair. At the same time, products should be designed to be remanufactured and ultimately to be recycled at the end of their life cycle. Thus, S.PSS design requires the integration of **product eco-design** (or **Life Cycle Design**), which focuses on reducing the environmental impact of a product looking at its different life cycle stages, from the extraction of raw materials, through manufacturing, distribution and use, and on to final disposal [24, 33, 38].

Looking more at the user-related aspects, it should be mentioned that some S.PSSs require a certain degree of change in patterns of consumption and user habits. Typically, this involves a shift from consumption based on ownership to consumption based on access and sharing goods. Even if we are generally used to not owning and sharing certain products (for example, of the products linked to mobility services like a car or bicycle), for some product categories (e.g. appliances) there are still substantial barriers for the adoption of S.PSS-oriented offers [35]. For this reason, it becomes crucial to design S.PSS offers able to stimulate changes in user behaviour and thus support the adoption of these kinds of solutions. **Design for sustainable behaviour** (e.g. see [15, 16]), and its ability to shape and influence human behaviour to support the adoption of sustainable innovations and habits, can thus play an important role in fostering the diffusion of S.PSSs. Design for sustainable behaviour can be applied to both the product and service elements of an S.PSS (e.g. services should be designed in a way that "sharing" should be seen positively throughout the user experience).

Emotionally durable design (e.g. see [8, 21]) can also be used to support S.PSSs. Emotionally durable design focuses on enhancing and strengthening the emotional tie between the user and the product so that the user–product relationship remains satisfactory over time. In those S.PSSs in which users have individual and long-term access to a product (e.g. product-lease) it might be beneficial to create a strong emotional connection between the user and the product, and thus adopt emotionally durable design strategies.

It is also important to note the potential linkages between S.PSS and social innovations. We must acknowledge that PSS design can take inspiration from social innovations to develop new product-service offerings (e.g. commercial vegetable box subscription services that mimic similar solutions developed at a local level by communities and farmers) [7]. Thus, **design for social innovation** (defined as "a constellation of design initiatives geared toward making social innovation more probable, effective, long-lasting, and apt to spread" [17]) can enable designers to gather inspirations from community-based solutions to ideate and develop S.PSSs. On the other hand, an S.PSS design can be used as an approach to foster social innovation by triggering, sustaining and/or guiding the direction of action. Finally, we need to highlight that S.PSS innovation can be complex to implement and bring to the mainstream, as they are hindered by a range of barriers [5, 20, 27, 35]: cultural barriers (e.g. the cultural shift necessary to value ownerless offers as opposed to owning products), corporate barriers (e.g. the need to implement changes in the business mindset and strategy) and regulative barriers (e.g. lack of internalization of the environmental and social costs in market prices). **Design for sustainability transitions**, which focuses on the transformation of socio-technical systems through technological, social, organizational and institutional innovations [7], can be used to support a successful implementation of S.PSSs. In particular, it can be adopted to understand the process of the introduction and diffusion of S.PSSs and how it can be more effectively designed, managed and oriented (e.g. see [3–5, 14, 34]).

At this point, it is also useful to discuss the relationship between S.PSS and the concept of the **circular economy**. The Ellen MacArthur Foundation [9] defined circular economy as "an industrial economy that is restorative or regenerative by intention and design". Its key principles are the creation of closed-loop systems of material flows and the 3R concept (reduction, reuse and recycling of resources)

[11]. As noted by Ceschin and Gaziulusoy [6, 7], even if the concept of the circular economy has been popularized and branded by Dame Ellen MacArthur, it can be considered as an umbrella concept that encompasses different principles that have been around for a long time (e.g. industrial ecology, biomimicry and cradle-to-cradle).

S.PSS design is crucial to support a circular economy; it can lead to business models that enable and foster circularity. As noted by Ceschin and Gaziulusoy [7], with the popularization of the circular economy concept, the term circular business model (e.g. see [22]) has gradually emerged. Bocken, Pauw, Bakker and van der Grinten [1] have proposed six circular business model strategies, grouped into two main categories:

- *strategies for slowing loops*, which include access and performance models, extending product value, classic long-life model and encouraging sufficiency; and
- *strategies for closing loops*, which include extending resource value and industrial symbiosis.

Apart from the different terminology and classification, the concept of S.PSS overlaps with the concept of circular business models. However, the circular business models include additional broader aspects, such as extending resource value (e.g. collection of otherwise 'wasted' materials/resources and turning them into new forms of value; [1]) and industrial symbiosis. Circular business models have a strong focus on the economic and environmental dimension of sustainability and less on the socio-economic dimension. In any case, it is clear how the concept of S.PSS represents a fundamental component of any circular economy. In fact, as stated in the recently published "EU Circular Economy Action Plan" [10], moving towards "product-as-a-service or other models where producers keep the ownership of the product or the responsibility for its performance throughout its lifecycle" is considered a key principle for sustainability, and a necessary condition to incentivize the design of sustainable products.

References

1. Bocken NMP, de Pauw I, Bakker C, van der Grinten B (2016) Product design and business model strategies for a circular economy. J Ind Prod Eng 33(5):308–320
2. Brezet H, Bijma AS, Ehrenfeld J, Silvester S (2001) The design of eco-efficient services: Methods, tools and review of the case study based "Designing Eco-efficient Services" project. Report for Dutch Ministries of Environment (VROM). VROM, The Hague, the Netherlands
3. Ceschin F (2012) The introduction and scaling up of sustainable product-service systems: a new role for strategic design for sustainability. Doctoral dissertation. Politecnico di Milano, Milan, Italy
4. Ceschin F (2013) Critical factors for implementing and diffusing sustainable product-service systems: Insights from innovation studies and companies' experiences. J Clean Prod 45:74–88
5. Ceschin F (2014) Sustainable product–service systems: between strategic design and transition studies. Springer, London, UK

6. Ceschin F, Gaziulusoy İ (2016) Evolution of design for sustainability: from product design to design for system innovations and transitions. Des Stud 47:118–163
7. Ceschin F, Gaziulusoy İ (2020) Design for sustainability: a multi-level framework from products to socio-technical systems. Routledge. https://doi.org/10.4324/9780429456510
8. Chapman J (2005) Emotionally durable design: objects, experiences, and empathy. Earthscan, London, UK
9. Ellen MacArthur Foundation (2013) Towards the circular economy: economic and business rationale for an accelerated transition, vol 1. Ellen MacArthur Foundation. https://www.ellenmacarthurfoundation.org/assets/downloads/publications/Ellen-MacArt hur-Foundation-Towards-the-Circular-Economy-vol.1.pdf
10. EU (European Union) (2020) Circular economy action plan: for a cleaner and more competitive Europe. European Commission, Brussels, Belgium
11. Geng Y, Doberstein B (2008) Developing the circular economy in China: challenges and opportunities for achieving 'leapfrog development'. Int J Sustain Dev World Ecol 15(3):231–239
12. Goedkoop M, van Halen C, te Riele H, Rommes P (1999) Product service systems, ecological and economic basics. Report 1999/ 36. VROM, The Hague, the Netherlands
13. Hockerts K, Weaver N (2002) Towards a theory of sustainable product service systems—what are the dependent and independent variables of S-PSS? In: Proceedings of the INSEAD-CMER research workshop "sustainable product service systems—key definitions and concepts", Fontainebleau, France, 9 May
14. Joore P, Brezet H (2015) A multilevel design model: the mutual relationship between product–service system development and societal change processes. J Clean Prod 97:92–105
15. Lilley D (2007) Designing for behavioural change: reducing the social impacts of product use through design. Doctoral dissertation. Loughborough University, Loughborough, UK
16. Lockton D, Harrison D, Stanton NA (2010) The design with Intent method: a design tool for influencing user behaviour. Appl Ergonom 41(3):382–392
17. Manzini E (2014) Making things happen: social innovation and design. Des Issues 30(1):57–66
18. Manzini E, Vezzoli C (2003) A strategic design approach to develop sustainable product service systems: examples taken from the 'environmentally friendly innovation' Italian prize. J Clean Prod 11(8):851–857
19. Meadows DH, Meadows DL, Randers J, Behrens WW (1972) Limits to growth. Universe Books, New York
20. Mont O (2002) Clarifying the concept of product–service system. J Clean Prod 10(3):237–245
21. Mugge R (2007) Product attachment. Doctoral dissertation. Delft University of Technology, Delft, the Netherlands
22. Nußholz JLK (2017) Circular business models: Defining a concept and framing an emerging research field. Sustainability 9(10):1810
23. Stahel WR (1997) The functional economy: Cultural and organisational change. In: Richards DJ (ed) The industrial green game: implications for environmental design and management. National Academy Press, Washington, DC
24. Tischner U, Charter M (2001) Sustainable product design. In: Charter M, Tischner U (eds) Sustainable solutions: developing products and services for the future. Greenleaf Publishing, Sheffield, UK, pp 118–138
25. Tischner U, Ryan C, Vezzoli C (2009) Product-Service Systems. In: Crul M, Diehl JC (eds) Design for sustainability (D4S): a step-by-step approach. Modules, United Nations Environment Programme (UNEP)
26. Tukker A (2004) Eight types of product-service system: eight ways to sustainability? Experiences from SusProNet. Bus Strategy Environ 13:246–260
27. Tukker A, Tischner U (eds) (2006) New business for Old Europe: product services, sustainability and competitiveness. Greenleaf Publishing, Sheffield, UK
28. UN General Assembly (2015) Transforming our world: the 2030 Agenda for Sustainable Development, 21 October 2015, A/RES/70/1. https://www.refworld.org/docid/57b6e3e44.html

29. UNEP (2002) Product-service systems and sustainability. Opportunities for sustainable solutions (ed) United Nations Environment Programme, Division of Technology Industry and Economics, Production and Consumption Branch, Paris
30. Van Halen C, Vezzoli C, Wimmer R (eds) (2005) Methodology for product service system innovation: How to develop clean, clever and competitive strategies in companies. Van Gorcum, Assen, the Netherlands
31. Vezzoli C (2007) System design for sustainability: theory, methods and tools for a sustainable "satisfaction- system" design. Maggioli Editore, Rimini, Italy
32. Vezzoli C (2010) System design for sustainability: a promising approach for low and middle-income contexts. International forum Design for Sustainability in Emerging Economies. June 2–5, 2010, Mexico-City, Autonomous Metropolitan University of Mexico City
33. Vezzoli C (2018) Design for environmental sustainability: Life cycle design of products. Springer, London, UK
34. Vezzoli C, Ceschin F (2008) Designing sustainable system innovation transitions for low-industrialized contexts. A transition path towards local-based and long lasting mobility solutions in African contexts. In SCORE!, (Sustainable Consumption Research Exchange!), Sustainable Consumption and Production: framework for action. Brussels, Belgium, 10–11 March 2008
35. Vezzoli C, Ceschin F, Diehl JC, Kohtala C (2015) New design challenges to widely implement 'Sustainable Product Service Systems. J Clean Prod 97:1–12
36. Vezzoli C, Ceschin F, Osanjo L, M'Rithaa MK, Moalosi R, Nakazibwe V, Diehl JC (2018) Designing sustainable energy for all. sustainable product-service system design applied to distributed renewable energy. Springer, London
37. Vezzoli C, Kohtala C, Srinivasan A, Diehl JC, Fusakul M, Xin L, Sateesh D (2014) Product-service system design for sustainability. Greenleaf Publishing, London
38. Vezzoli C, Manzini E (2008) Design for environmental sustainability. Springer, London, UK

Distributed Economies

Aguinaldo dos Santos, Carlo Vezzoli, Brenda Garcia Parra,
Sandra Molina Mata, Sharmistha Banerjee, Cindy Kohtala,
Fabrizio Ceschin, Aine Petrulaityte, Gabriela Garcez Duarte,
Isadora Burmeister Dickie, Ranjani Balasubramanian, and Nan Xia

1 Reframing the Economy Towards Sustainability

There is an urgent need to reframe the economy towards a new paradigm where economic evolution occurs fairly and ethically, in conjunction with the development of human well-being achieved in harmony with nature. This emerging paradigm presents profound divergences from the orthodox paradigm, which is based on economic rationality (characterized by a continuous pursuit of economic efficiency

A. dos Santos (✉) · G. G. Duarte
Universidade Federal do Paraná, Curitiba, Brazil
e-mail: asantos@ufpr.br

C. Vezzoli
Design Department, Politecnico di Milano, Milan, Italy

B. Garcia Parra · S. Molina Mata
Universidad Autónoma Metropolitana, Mexico City, Mexico

S. Banerjee
Indian Institute of Technology Guwahati, Guwahati, India

C. Kohtala
Department of Design, Aalto University School of Arts, Design and Architecture, Espoo, Finland

F. Ceschin · A. Petrulaityte
Department of Design, Brunel University London, London, UK

I. B. Dickie
Universidade da Região de Joinville (Univille), Joinville, Brazil

R. Balasubramanian
Srishti Institute of Art, Design and Technology, Bengaluru, India

N. Xia
Tsinghua University, Beijing, China

© The Author(s) 2021
C. Vezzoli et al. (eds.), *Designing Sustainability for All*, Lecture Notes
in Mechanical Engineering, https://doi.org/10.1007/978-3-030-66300-1_2

Table 1 Comparing two economic paradigms [57]

Orthodox paradigm	Promising sustainability paradigm
Individualism	Solidarity
Growth	Development
Large scale	Small scale
Competition	Cooperation
Centralization	Distribution
Profit	Well-being
Tangible	Intangible
Product based	Service based
Reduced ethics	Ethical and fair
Consumerism	Sharing

in resource exploitation) [13, 17, 39, 55, 60, 66]. In a sustainable approach, solutions should jointly promote the improvement of welfare, social cohesion and social equity, while significantly reducing environmental impact and resource depletion. Table 1 illustrates the main differences between these economic paradigms.

This new economic paradigm includes cooperative work in the production of goods and services, solidarity finance, fair trade and solidarity consumption (MTE 2012). An initiative or enterprise is guided by the generation of work and income and, at the same time, seeks to achieve social inclusion and respect for ecosystems. The economic, political and cultural results obtained from value creation are shared among participants, thus constituting a strategy to overcome the pattern of subordination and vulnerability observed in conventional practices prevalent in the orthodox economy [27]. The implementation of such a vision has the excessive centralization of the economy as one of its key barriers, as explained in the next section.

2 How Centralization Hinders Sustainability and Resilience

The rationale for centralizing, mass production for economies of scale has been based on the ideals of efficiency and cost-savings, ideals that are rarely tested for their real efficiency or efficacy [11, 31, 43]. For example, in the case of electricity, a certain percentage is always lost in transmission, particularly in grids that are not well maintained. Manufacturing of goods by centralized mass production becomes efficient particularly when the social and environmental costs of manufacturing, from waste and pollution to decent working conditions, are externalized. In the worst case, overcapacity may be pushed onto consumers through aggressive marketing as well as planned obsolescence strategies, and nature is seen only a provider of 'resources', raw materials and raw land to be exploited. Much critique of current industrial mass production thus centres on tendencies to promote consumerist values, overconsumption and throwaway products. Large firms are also less likely to answer to consumer pressure for environmental and social responsibility; simultaneously, the large distances between consumers and manufacturing supply chains means consumers are not always fully aware of sustainability issues [5].

In the fast fashion industry, for instance, production efficiency, low wages and dangerous working conditions in many regions have radically reduced the prices of apparel for consumers, which in turn has increased consumption—and its negative social and environmental impacts—by even 40% [46]. Negative environmental impacts from the production of fast fashion include substantial use of water and chemical pollutants, especially in regions of water scarcity and less capacity for environmental protection measures, not to mention impacts from transportation, retail distribution and disposal [46]. Such impacts are usually experienced in low- and middle-income regions far from where the clothing is purchased. Moreover, there are negative environmental impacts from waste in many industries, which includes not only pre-consumer waste produced during manufacturing, but also "deadstock"—finished goods such as fast fashion and luxury goods that are disposed of before they even reach the consumer [45]. Deadstock is surplus output, a direct result of overproduction in centralized, large-capacity, capital-intensive mass production, in contrast to other models such as production-on-demand. The principle of Distributed Economies therefore calls for an analysis of what products and services in a specific region deliver social and environmental harms by virtue of being produced in large-scale, centralized modes. The objective is to become sensitive to and work to change systems that have become an "ever-faster once-through flow of materials from depletion to pollution" [11, 28].

Critique of 'centralization' is not limited to tangible products and their manufacturing. In the fast fashion example, attention is also paid to what consumer behaviours are encouraged as a result of low prices in a consumerist society, which impose barriers to other experimental models such as sharing, renting and upcycling that would extend product and material lifetimes. These alternative models also connect actors in other ways than fiat money, connections that are not visible or valued in models that emphasize capital-intensive, efficient, centralized industrial systems [38, 44, 54] . At the same time, one must also be wary of centralizing tendencies in the "sharing economy". As the largest peer-to-peer platforms for "collaborative consumption" gain critical mass, while retaining ownership in centralized corporate hands far from local users, there is uncertainty and controversy over how such "platform capitalism" delivers social benefits, local value and positive environmental impacts for their diverse stakeholders [20, 44, 53, 59].

Another critique of institutional centralizing relevant to design relates to expertise and legitimacy: who has the authority to produce, design, innovate and distribute. Centralized production, geographically and/or via patents and Intellectual Property Rights regimes, separates the authority to repair and maintain from the knowledge to repair and maintain, for instance. Such barriers can affect actors who contribute to a local economy and ensure circular material flows (through product longevity)—such as repair hackerspaces—but are not accounted for in neoclassical economics indicators [31, 54]. Analysis according to DE principles would therefore examine where economic activities threaten local resilience and the ability to satisfy local needs, cases where "one industrial production process exercises an exclusive control over the satisfaction of a pressing need and excludes nonindustrial activities from competition" [29]. It is beneficial for environmental, social and economic sustainability that knowledge of design and abilities to innovate are not removed from communities, but are rather enhanced.

Centralized systems and the accompanying extreme focus on efficiency must thereby be examined in terms of sustainability because of the impact on societies' and systems' resilience. If resilience is understood as the ability for a system (such as a city, region or neighbourhood, including natural systems and industrial systems) to be flexible, agile, adaptive and able to absorb shocks from a disturbance [24, 26, 50], an excessive focus on efficiency leads to structures that are fragile and brittle [50]. Both ecosystems and human systems absorb shocks and deal with disturbances by "allowing the existence of some redundant and not-so-efficient pathways" [9, 50]. From the point of view of a city, resilience would address dependence on global supply networks and the need for diversified economic activities, which requires examination of the role of mass manufacturing and services in the region [24].

The shift to a network society [12] appears to embed new potential: new ways societies can meet their needs and express themselves creatively, which call into question—and actively dismantle—harmful systems [6]. Walter Stahel suggests shifting emphasis from production optimization to use optimization, and that large-scale, capital-intensive production units be gradually or partially replaced by "smaller-scale labour-intensive, independent, locally integrated work units" [62]. This "distributed" model is the focus of the next section.

3   Distributed Economies (DE) as a Strategy Towards Sustainability

Distributed Economies consists of small-scale value-adding units (e.g. manufacturing, services) where there is a shift in the control of core activities towards the user/client. Johansson et al. [31] first defined Distributed Economies as a "selective share of production distributed to regions where activities are organized in the form of small scale, flexible units that are synergistically connected with each other" in a network.

These local units serve local needs near or at the point of use, including artefact and service demands across the product life cycle and business process, shifting the control of essential activities towards or by the end-user, whether individuals, entrepreneurs or organizations. Hence, in such contexts, local units are more capable of offering on-demand solutions and having a higher level of multi-user participation, including those situations where the user her/himself can also take the role of manufacturer or service provider.

In a Distributed Economy, these small-scale units could be stand-alone or peer-to-peer, connected with other nearby units to share various forms of products, semi-finished products, resources, knowledge/information and other types of services. These local units are sometimes organized as multiple providers to the same order, forming a much more resilient network (e.g. cooperatives). Hence, this local network can be connected to nearby networks, resulting in an expanded network of networks, i.e. they become a Distributed Economy Network (DEN). If properly designed taking sustainability principles into account, they have potential to promote locally based sustainability, i.e. Sustainable Distributed Economies (S.DE). They share or jointly use various forms of local resources, including skills, knowledge and manufacturing/service capabilities.

When we discuss the concept of Distributed Economies, we do so in contrast to Centralized Economies for simplicity and clarity in analysis. With that in mind, we can identify two types of small-scale locally-based production units where we find a shift in the control of core activities towards the user/client. The first we (also) call *Distributed*, which are *by the end-user*, and the second *Decentralized*, which are *nearby the end-user*, as illustrated in the diagram below (Fig. 1).[1]

In contrast to DE, a **Centralized Economy** is *characterized by* **large production units** *located (often) far from its customers (individuals or organizations), with production capacity geographically concentrated, delivering products/services* via **large distribution networks**. *Their large-scale, stand-alone production units demand high control of essential activities and, thus, decision making is often centralized.* Due to their scale, implementation of changes is often costly and time-consuming (Fig. 2).

Meanwhile, a **Decentralized Economy** is *characterized by* **small**-*scale* **production units** *that deliver their goods and services* via **light distribution** *networks, directly to* **customers,** *whether individuals, entrepreneurs or other organizations/institutions, increasing* **customers' control over** *essential* **activities;** *they could be* **stand-alone** *or* **connected** *to each other to* **share** *various forms of* **goods** *and* **services**. *Thus, the cost and time for implementing or changing them is also variable. Their decision-making process is decentralized, with some customer/user control over essential activities* (Fig. 3).

Finally, a **Distributed Economy** *involves (very) small*-*scale* **production units** *of goods (physical and/or knowledge-based artefacts) located* **near or at the same place of the end-users** *(who become the producers, i.e. prosumers) that have* **control over** *essential* **activities**, *whether individuals, entrepreneurs or organizations/institutions. They could be* **stand-alone** *or peer-to-peer* **connected** *to each other to* **share** *various forms of* **goods** *and* **services** (see Fig. 4).

A Distributed Economy (DE) could be further characterized by its life cycle localization depth, i.e. whether it is centralized, decentralized or distributed along all its life cycle stages (pre-production, production, distribution, use and disposal). The relevance and configuration of these stages could differ from case to case, as exemplified in the right-most diagram in Fig. 5, which describes the life cycle localization

---

[1] We thereby use this terminology and conceptualization in this volume, acknowledging that these terms have different definitions in various fields.

depth of a solar panel produced and distributed by a multinational company (Centralized), installed and used by an individual, e.g. having it installed on the roof of their home (Distributed), and disposed of locally (Decentralized). An in-depth analysis of this example shows the system is Centralized in its pre-production, production and distribution phase, Distributed in its use phase and Decentralized in its disposal phase.

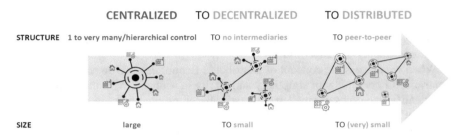

Fig. 1 The paradigm shift from centralized, to decentralized, to distributed economies

Fig. 2 The structure of the production unit of Centralized Economies

Fig. 3 The structure of the production unit of Decentralized Economies

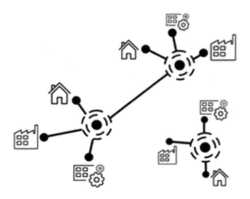

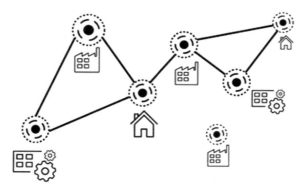

Fig. 4 The structure of the production unit of Distributed Economies

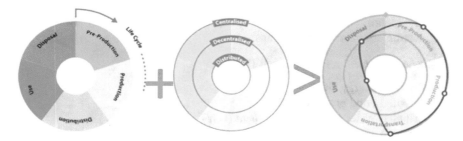

Fig. 5 Example of how a system should be characterized according to how Distributed/Decentralized/Centralized it is in its various life cycle stages

When compared to the Centralized approach, Distributed Economies is a promising offer model for enhancing cohesion to the same goals and more equitable distribution of power at a local level, distributing the activities based on expertise, resource availability and accessibility. Furthermore, these flexible units may have less emphasis on economic growth and more on the achievement of well-being. Therefore, its adoption implies a rupture to the unsustainable foundations of neoclassical economics, which is often driven by the idea that large-scale production makes better economic sense.

It is useful to observe, furthermore, that Distributed Economies (DE) is nothing new. What we have experienced over the course of decades has been and is a process of centralization, especially in industrialized countries. For example, preparing a meal at home is a distributed activity with its home-based production units (ovens, etc.). Nevertheless, even in this case, we may observe an evolution towards a life cycle centralization (in industrialized and emerging contexts): to cook we buy electricity/gas from the main grid (centralized), while in the past we collected nearby biomass (distributed, though with highly toxic combustion fumes); we can purchase food in a supermarket (centralized), while in the past much was cultivated in our gardens or bought from neighbourhood shops (distributed); finally, nowadays we can buy "almost ready meals" (centralizing most cooking activities).

Furthermore, DE already exists in many low- and middle-income contexts. In Kenya, for instance, according to the International Labor Organization, 90% of businesses are informal, which would mean that a large percentage of the population is familiar with the distributed and networked nature of the informal sector. Such a population could already be familiar with the open and networked relationships that S.PSS and DE offer [67]. Hence, Sustainable Distributed Economies (S.DE) need to be seen not as a return to the past, but as a transition towards socially, environmentally, economically and technically advanced sustainable distributed economies.

4 Practical Implications of DE in Various Fields

We may identify different types of Distributed Economies (DE). Below is a classification organized in two groups:

Hardware/natural resource-based DE:

- Distributed energy Generation (DG),
- Distributed Food production (DF),
- Distributed Water supply/management (DW)
- Distributed Manufacturing (DM).

Knowledge/information-based DE:

- Distributed Software development (DS),
- Distributed Knowledge generation (DK),
- Distributed Design (DD).

These DE types are described in the following sections.

4.1 *Distributed Design (DD)*

A Distributed Design (DD) system is an open design system where solutions are conceived and/or developed by a small-scale design unit, e.g. one person/computer being the end-user or located nearby the end-users, whether individuals, entrepreneurs and/or organizations/institutions. If the small-scale production units are also connected with other DD (e.g. to share the open design technical drawings), they become a Distributed Design Network (DDN), which may in turn be connected with nearby, similar networks. If properly designed, they are promising to promote locally based sustainability, i.e. Sustainable Distributed Design (S.DD) systems. Through participatory design practices in the context of digital technologies, such as open design and crowd-design [19], designers can access widely dispersed or demographically segmented user groups and suppliers, engaging them directly to contribute with ideas and solutions, and encouraging them to engage in the outcome configuration. In this way, the development of a new product, service or Product-Service System can be done by laypeople, prosumers, producers, creative communities, experts in various fields, designers and companies, or even by the interaction between these groups [18]. The collaboration between the people involved in the development of these projects can occur through crowd-based platforms, FabLabs, makerspaces, hackerspaces, or iteratively between these spaces [15].

4.2 Distributed Manufacturing

Distributed Manufacturing (DM) can be described as a production system made of small-scale manufacturing units equipped with physical and digital technologies, which enable the localization of manufacturing facilities and comprehensive communication between all supply chain actors in order to facilitate customer-oriented production [49]. Key DM features can be summarized into three categories: the localization of manufacturing units, the application of physical and digital technologies, and the customer orientation [7, 34, 61]. The localization of manufacturing units addresses the proximity between manufacturing facilities (e.g. factories, workshops, personal fabrication labs or makerspaces, in-house and in-store suites, mobile manufacturing units, etc.) and end customers and/or manufacturing resources. The application of physical and digital technologies refers to hardware, tangible manufacturing equipment needed to produce products (e.g. 3D printers, laser cutters, Computer Numerical Control (CNC) routers) and the application of computer systems and the use of the Internet (e.g. Industry 4.0, Cloud Computing, Internet of Things, ICT, etc.) used to collect and process data and enable communication between key actors. The customer orientation refers to the level of product or service customization (e.g. mass customization, personalization, bespoke production, etc.) and the level of customer involvement in design and production processes.

Implementation of DM brings multiple benefits for companies and their customers, including companies' resilience to changes in market demand [51], enablement of personalized production [32], facilitated movement and relocation of manufacturing facilities [61], reduction of supply chain actors [4], and many more. However, the transition towards DM requires companies to change organizational mindset [8], adopt new ways of managing business processes [47] and invest in new manufacturing and communication technologies [4].

4.3 Distributed Energy Generation (DG)

Decentralized and distributed energy generation systems (DG systems) are typically powered by renewable energy sources. These include solar, wind, small hydro, biomass, biogas and geothermal power.

There is no consensus on a shared definition of decentralized generation and distributed generation [22]. For some authors, these two terms are synonymous [33]. For others, the difference is that in decentralized systems, the energy generation units have no interactions with each other [2, 36]. At any rate, from a technical perspective we can distinguish between [21, 65]:

(a) Stand-alone energy systems: these are off-grid systems, thus not connected to each other or to the main grid;
(b) Grid-based systems: these are energy generation systems which supply power at a local level, using local-wide distribution networks [52].

DG systems are associated with a range of potential sustainability benefits [21, 65]:

- From the economic perspective, DG systems are characterized by lower transmission costs for remote regions and lower energy prices in the long-term compared to centralized systems [48]. They can also enhance the flexibility and resilience of the system [31]. A system can easily cope with individual failures (i.e. fault in an energy generation unit) since each energy-using node can be served by multiple energy production units. DG systems require relatively low investments, making it easier for small economic entities such as single individuals and/or local communities to become prosumers (consumers but also producers of the energy).
- In relation to the environmental aspects, the use of renewable and locally available energy sources results in a lower environmental impact compared to the use of fossil fuels (and the related extraction, transformation and distribution processes) [58]. Moreover, local energy production reduces the energy distribution losses that characterize centralized systems.
- Regarding the socio-ethical dimension, the fact that DG systems are relatively easy to be installed and managed (and thus enable users to become prosumers) fosters the process of democratization of energy access, thus enhancing community self-sufficiency and self-governance [14]. Additionally, being locally distributed, they can lead to an increase in local employment (e.g. in relation to installation and maintenance activities) and thus dissemination of competences, which can foster local economies.

However, despite their potential benefits, there are also some barriers to be taken into consideration (for a more detailed discussion see [65], Sect. 5): technical (e.g. resource availability, skill requirement for design and development), economic (e.g. users' purchasing power and spending priorities, energy pricing, incentives), institutional (e.g. policy and regulations), socio-cultural (e.g. norms and value system, behavioural or lifestyle issues), and environmental (e.g. impact on ecosystems and wildlife).

An example of Distributed Energy Generation is the solution offered by IBEKA, a non-profit organization operating in Indonesia. IBEKA provides hydro mini-grids to communities. This includes the design and installation of the energy generation plant as well as support to enable the local community to manage the plant. IBEKA sets up a community-managed enterprise to run the system and trains it on how to operate, maintain and manage it. The grid-connected system allows communities to sell surplus energy to the national energy supplier. Revenues cover operation, maintenance, loan repayments and a community fund. End-users pay according to a tariff which could be based on a pay-per-energy consumed (meter) or an agreed amount of energy per day.

4.4 Distributed Water Supply/Management

A Distributed management system of Water (DW) is a small-scale management unit, located by or nearby the end-users, whether individuals, entrepreneurs and/or orga-nizations/institutions. If the small-scale Water supply/management unit (DW) is also connected with other DWs (e.g. to share the water surplus), they become a Distributed Water supply/management Network (DWN), which may, in turn, be connected with similar networks nearby. If properly designed, they have potential to promote locally based sustainability, i.e. Sustainable Distributed Water supply/management (S.DW) systems. An example of a Distributed Water supply/management (DW) system is the shift from a centralized urban water supply to distributed access to clean groundwater.

Compared with water supply/management systems based on centralized systems, distributed systems are smaller in scale. In structure, the relationship between produc-tion units is more equal. It is also more flexible and proactive; compared to the central type, the production unit of the distributed system is closer to the user and more open, which can motivate users to actively participate and develop customized solutions to effectively meet individual needs [68].

For example, P1MC is a charity project initiated by the Brazilian NGO ASA in early 2000 to help residents of the arid regions of north-eastern Brazil to build home rainwater storage facilities. P1MC abandoned the traditional water tank product sales model, but supported local villagers to build their own reservoirs, provided training on routine maintenance methods and provided follow-up technical support. This model of 'collaborative construction' plus 'services and training' has a significant role in promoting project implementation in poor areas. Through professional planning and design, local organizations are encouraged to collaborate with individuals, signifi-cantly reducing the cost of building and operating hardware facilities and making local water supply solutions more flexible and agile.

4.5 Distributed Food Production (DF)

Distributed Food production (DF) is a small-scale value-added unit (production/service) associated with food, located by or nearby the end-users, whether individuals or organizations. If the small-scale Food production units (DF) are also connected with other DF (e.g. to share food overproduction), they become a Distributed Food production Network (DFN), which may, in turn, be connected with similar networks nearby. If properly designed they have potential to promote locally based sustainability, i.e. Sustainable Distributed Food production (S.DF) systems.

Centralized food systems evolved along with the advances of the industrial revolution, adopting production and consumption practices based on industrial, mass production logic, that is, introducing elements that aim for system optimization and production efficiency, prioritizing financial gain over quality of food produced. In a period of little more than 200 years, in order to guarantee the expansion of the agricultural frontier and the volume of food production, agro-industrial practices have progressively been adopting mechanization, introducing chemical substances and promoting genetic modification as support pillars of the system. This has put the survival of millenary practices and traditions that revolve around food at risk, without taking into consideration the impact of such practices on the natural and social systems that sustain it, resulting in the consequent socio-environmental degradation of the planet.

Alternatives as Distributed Food production encompass a comprehensive set of ideas that have put into practice the diffusion of community networks and the quest for small-scale and flexible sustainable solutions, making use of local resources. Initiatives include Experiential Agribusiness, Community Supported Agriculture, Urban Farming and the Slow Food movement.

Experiential Agribusiness is based on the offer of gastronomic experiences as a value proposition. It can be considered a decentralized, small-scale system that appropriates traditional food production techniques and cultural practices, reconfiguring new gastronomic propositions strongly influenced by user experience under the name of food design. Community Supported Agriculture focuses on the production of high-quality foods for a local community, often using organic or biodynamic farming methods and a Decentralized or Distributed structure. It connects the producer and consumers within the food system by allowing the consumer to get involved in the different activities related to the harvest of a certain farm or group of farms. Urban farming is the practice of cultivating, processing and distributing food and the raising of animals for food and other uses within and around cities and towns. It takes advantage of vacant and underutilized private or public spaces within the city and the suburbs that might have a potential use for farming purposes. Slow Food is a global movement present in more than 150 countries. It is a reference in debates on biodiversity, local food communities and genetically modified food [3]. It was initiated with the aim to protect regional traditions, good food, gastronomic pleasure and a slow pace of life from the perceived domination of agribusiness, supermarkets and fast food chains.

4.6 Distributed Software Development (DS)

Distributed Software development (DS) is a small-scale production unit (i.e. a computer is the basic hardware for such production), located by or nearby the end-users, whether individuals, entrepreneurs and/or organizations/institutions. If the DS small-scale production units is also connected with other DS (e.g. to share information, open data or open code), they become a Distributed Software Network (DSN), which may, in turn, be connected with similar networks. If properly designed, they hold promise to promote locally based sustainability, i.e. Sustainable Distributed Software development (S.DS) systems. A well-known example of a Distributed production of Software (DS) is the shift from proprietary software to open-source software 'Linux'.

4.7 Distributed Production of Knowledge (DK)

A Distributed production of Information/Knowledge (DK) system is a small-scale production unit (i.e. a computer is the basic hardware for such production), located by the end-users or peer-to-peer connected with the end-users, whether individuals, entrepreneurs and/or organizations/institutions. If the DK small-scale production unit is also connected with other DK (for example, to share open information and data), they become a Distributed Knowledge generation Network (DKN), which may, in turn, be connected with similar networks nearby. If properly designed, they hold promise to promote sustainability on a multilocal level, i.e. Sustainable Distributed Knowledge generation (S.DK) systems. A well-known example of Distributed Information/Knowledge generation is the shift from the traditional encyclopaedia to the open encyclopaedia 'Wikipedia'. In fact, the LeNS Learning Network on Sustainability of HEIs could be classified into this category.

5 Alternative System Configurations

5.1 Stand-Alone Configurations

A stand-alone DE configuration occurs in those systems characterized by the use of either distributed or decentralized production units, without any local delivery system (network) with nearby customers and/or production units. These isolated production units are run by and for the user, either by an individual or an enterprise/organization. A Stand-Alone Distributed system is an isolated production unit by the end-user, while a Stand-Alone Decentralized System is an isolated production unit reached by near-by customers to benefit from the outcomes (of the production unit) (see Fig. 6 below).

5.2 Network Configurations

There are four types of Network Configuration, as described below:

- A *Centralized Network System* is a network of production units far from the user with an extensive delivery system for various forms of resources (physical and/or knowledge-based) to individuals or enterprises/organizations distributed in a large-scale area such as a state/s, country/ies, continent/s or worldwide (see Fig. 6 below).
- A *Decentralized Network System* is production with a local delivery system (network) for various forms of resources (physical and/or knowledge-based) to nearby individuals or nearby enterprises/organizations (Fig. 6).
- A *Distributed Network System* is a network of production units run by the user, either an individual or an enterprise/organization (Fig. 6), sharing various forms of resources (physical and/or knowledge-based) locally with nearby individuals and/or organizations.
- A *Hybrid network* system is a network of production units that consists of two or more types of centralized, decentralized or distributed network systems (Fig. 6).

Beyond these four configurations, there can also be a **Network of Networks**, which are either centralized, distributed or decentralized production units or local networks connected to other networks to share various forms of resources (physical and/or knowledge-based) (Fig. 6).

Finally, a **DE can also be connected to a Centralized Network**. In this case, either distributed or decentralized production units or local networks are connected to a Centralized Network to share various forms of resources (physical and/or knowledge-based) (Fig. 6).

5.3 Summary and Examples of System Configurations

Figure 6 visually summarizes the main system configurations described in the previous section.

The following table gives examples from the different DE classifications for these alternative system configurations (Table 2).

6 Main Drivers and Win-Win Benefits of DE

Table 3 presents a wide range of win-win benefits of DE according to the three dimensions of sustainability [56]. Changes in customer behaviour and demands, including the quest for greater well-being and more democratic practices, are opening opportunities for a wider adoption of Distributed Economies. The proximity between

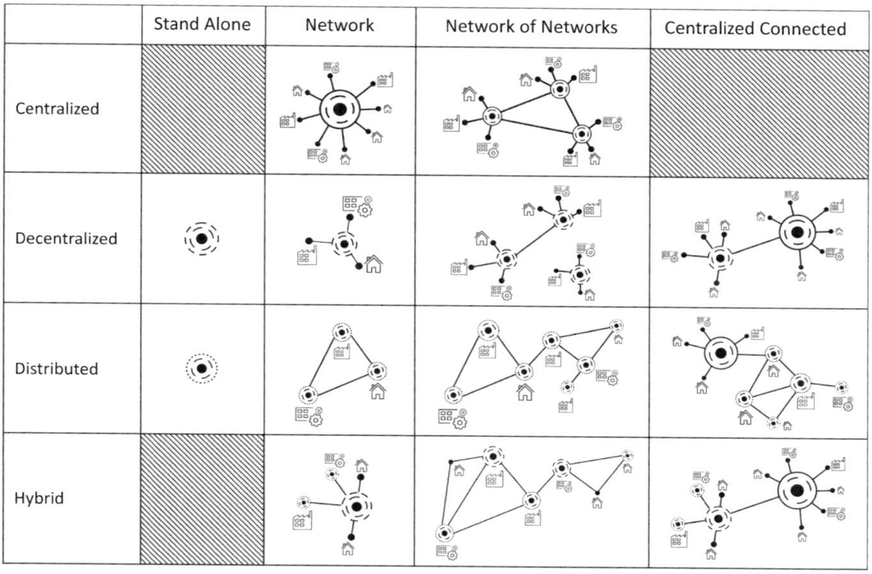

	Stand Alone	Network	Network of Networks	Centralized Connected
Centralized				
Decentralized				
Distributed				
Hybrid				

Fig. 6 Possible production/delivery system configurations

producers and consumers enables the provision of solutions with a better fit to local needs. By re-connecting people and producers, Distributed Economies also provide an opportunity for poverty alleviation, with people providing for their needs in alternative ways. Various authors [16, 31] argue that DE offers advantages in the pursuit of social diversity, respect for local culture, increased local quality of life and collective spirit, and focus on regional assets expanding the bargaining power for local actors beyond the maximization of social capital.

Some of the main economic drivers to adopt DE characteristics include the growing interest in customization and the reduction of logistics, lead time and labour costs due to shorter distances. In addition, the embedded characteristics of DE enable more collaborative design and production, with optimal distribution and use of resources. It is aligned to the expectations of a young generation that is increasingly in search for jobs with more freedom and creativity.

Emerging technologies have also opened new avenues and opportunities to implement DE. The possibilities provided by technologies such as IoT (Internet of Things), AI (Artificial Intelligence) and digital fabrication (such as Additive Manufacturing technologies), have aligned with a growing level of internet access and broader options for communication technologies. This has opened new avenues for merging digital and physical technologies, resulting in more flexible and agile manufacturing/services as well as knowledge sharing approaches.

Table 2 Examples of the main DE structure configurations

	Stand-alone—distributed	Stand-alone—decentralized	Network—distributed	Network—decentralized
DG (Distributed Energy Generation)	Solar panel for single household energy production	Hydro-powered charging station, where people go to charge phones, etc.	Solar panels for home use, connected via local mini-grid, sharing an energy surplus	Wind farm which provides energy to a village with a local mini-grid
DF (Distributed Food)	Home gardening for private use	Organic producer selling food directly to local consumers near or from the fields	Neighbourhood gardening club sharing of production surplus	Local baker delivering organic bread to the neighbours every morning
DW (Distributed Water)	Rainwater harvesting from the home roof for private use	Medium-sized water collector that local people access with their tanks to get the water	Roof rainwater harvesting for private use, with neighbourhood piping infrastructure for surplus sharing	Water from a local spring distributed to the households in the village through local infrastructure
DM (Distributed Manufacturing)	An individual making clothes at home using sewing machines for own use	A maker selling 3D printed artefacts directly to the final user in a shop beside the workshop	A digital fabricator supplying own needs while producing and delivering to locals during unused time	An entrepreneur locally delivering 3D printed items made on request
DS (Distributed Software)	A developer developing software at home to create a home security system	A software developer team selling the security system they developed from their office to local enterprises/organizations	A local community of developers collaborating on open-source software to create and install a home security system	A software developer providing a service installing the home security system she/he developed
DK (Distributed Knowledge)	A (very) small weather station for home forecast	A farming expert providing a consultancy service in her office about farming for the region	A small weather station located at an individual's home for their own use and sharing the data with the local community	A local consultancy providing a gardening service by going to the customers' gardens

(continued)

Table 2 (continued)

	Stand-alone—distributed	Stand-alone—decentralized	Network—distributed	Network—decentralized
DD (Distributed Design)	An individual designing his/her family clothes at home	A studio providing an onsite service for the local customers to design custom furniture	An individual designing their own clothes at home and sharing the designs with their local community	An architect studio providing services for a local community by going to their houses

Table 3 Main Win-Win Benefits of DE

Social benefits	Environmental benefits	Economic benefits
• Fosters a culture of mutual help and empowerment, enhancing the social resilience of the system; • Fosters higher socio-economic equity, offering more opportunity to marginalized people, thus accepting diversity; • Encourages the sharing of knowledge and skills, providing a better environment for wide competence building; • Values local culture, knowledge and capabilities by using local skills and native knowledge, enabling higher customer/user involvement in the design process; • Promotes social cohesion among local stakeholders, with a better cultural fit of products/services, creating meaningful and long-lasting relationships with customers, promoting mutual trust at the local level.	• Enables a shift towards a circular economy, making easier the collection of products at the end of their life cycle due to shorter distances; • Reduces environmental impact due to shorter distances, increasing system efficiency, with a decrease in the demand for resources and, at the same time, more emphasis on the use of renewable resources and conserving resources; • Increases the possibility to prioritize the environment over pure financial gains as users/clients can keep direct contact with the environmental impacts resulting from their choices; • Delivers a higher rate of shared services and resources, leading to better resource use and democratization of access to resources.	• Enables better fulfilment of local needs, allowing on-demand production and reduction of marketing costs due to customer proximity; • Provides a higher level of customization and enables faster delivery of product/service changes; • Features shorter, more flexible and smaller supply chains, with sharp reduction in logistics costs, lead-time, waste and capital investment; • Enables better monitoring of product performance, with higher local control over production; • Valorizes the local economy, integrates local competencies and infrastructure into the design process, increasing the bargaining power of local providers and encouraging open source innovation.

7 Potential Unsustainability of DE

'Distributed' does not automatically mean 'good' or anti-centralized, and these concerns are immediately apparent in the most extensive online peer-to-peer platforms, from sharing of services to social media [38, 40, 54, 59]. Even when people are seen as 'members of communities' socially connected to each other (compared to being mere providers of physical labour in a factory), they have nevertheless become providers of data that is sold by centralized media giants to other parties for profit. Individuals acting within these platforms do not become part of collective local economies, nor is their resilience necessarily enhanced by their participation.

As the notion of Distributed Economies emerged from Lund University's International Institute for Industrial Environmental Economics in the mid 2000s, IIIEE publications from that time have helped clarify what it is we *do not want* in our current global mass production-consumption system by emphasizing DE [31, 41]. The negative characteristics of 'centralization' discussed in this literature still hold true for products, services or platforms that appear to be decentralized, distributed and peer-to-peer, and analyses must account for this. Table 4 summarizes these and other main potential unsustainabilities of DE.

The shift to a network society has not yet been accompanied by a generalized knowledge of how to govern ourselves in horizontal networks that embed market-oriented, public-sector-oriented and civic-society oriented actors and actions—particularly when trying to keep ecological impacts in mind. Decentralizing and distributing too easily ends up as more business-as-usual. "[L]ocal actors' possibilities to have ownership and control over their immediate economic environment" may be strengthened in appearance, while weakened in operation [41]. It is thus essential to pay attention to what remains centralized, when limited conceptions of market value predominate, and when discussions on the nature of economic collaboration is depoliticized. Communities that strive to repoliticize the discussion on decentralizing, from Transition Towns to indigenous land defenders to open design groups working on sustainable solutions, make visible what is 'centralized' and why it is undesirable, and they actively prototype and prefigure new modes of production. By examining their examples, and how they interplay with mass production and consumption from the 'orthodox economy' (see Sect. 1), we see that characteristics such as standardization and modularity, for instance, are still useful, but useful for community autonomy and resilience, not for financial profit for a selected few.

Table 4 Potential unsustainabilities of DE

Potential environmental unsustainabilities	Potential social unsustainabilities
Large-scale centralized production units could optimize resource consumption and emissions (per production outcome)	DE production units are not necessarily empowering local economies and well-being
In centralized production units, labour practices could be more specialized ("expert"), i.e. optimizing resource consumption and emissions (per production outcome)	DE production units, particularly the increase of do-it-yourself, could at the same time decrease employment, as far as doing something by oneself reduces the opportunities to employ local experts
DE production units are not necessarily (designed) with a low environmental impact (e.g. to use renewable resources)	DE production units could be used merely as a strategy to outsource locally, without proper care for safety standards and the quality of life in workplaces
DE outcomes do not necessarily have a low environmental impact	An increase in the amount of local production or services may jeopardize social habits or routines
DE practices that involve increased digitalization may contribute to greater volumes of e-waste, increased electricity consumption, greater embodied energy of electronic system components and increased consumption of scarce resources such as rare earth metals	Local production or services may require expert knowledge and/or material or cultural resources that are not locally available

Especially in the last five to ten years, internet-enabled, open, peer-to-peer connectedness has enhanced our ability to participate and radically distribute tasks and activities. However, it has also weakened our physical and mental health, accelerated throughput of e-waste, increased our global need for energy, further marginalized the already marginalized, and threatened our very democracies. It appears, then, that we need to not only re-visit the literature but continually update our alternative conceptualizations of the economy and its role in structuring our relationship to the living earth and webs of life. For more resilient communities, the DE concept has emphasized good environmental performance, local people's preferences, quality of life and well-being [30], while particularly examining privileged regions in northern Europe. The Stockholm Resilience Centre has emphasized how humans and nature are intertwined in complex social-ecological systems, where resilience-building needs to nurture diversity, combine different types of knowledge for learning and create opportunities for self-organization [26], while remaining within the paradigm of 'development'. From the perspective of post-development and post-coloniality, acknowledging that global inequities are only increasing, Escobar [23] and others have emphasized plurality, community autonomy and self-determination.

To conclude, despite its potential unsustainabilities, DE still stands as a useful framework for understanding how we want to shape our local economies, even within a rapidly transforming, global environment with many industrial and post-industrial trajectories.

8   Understanding DE from Different Contexts

8.1   A Brazilian Perspective

The service sector is the largest component (70%) of the Brazilian national Gross Domestic Product (GDP). However, there is an uneven development pattern of the sector across the country. Service activities are concentrated in the same large poles with a North-South divide: the South concentrates the most dynamic sectors and providing greater diversity of services, as well as larger sizes of firm, i.e., greater economies of scale. The North, particularly in the northeast region, shows lower diversification of services and an intense concentration of the 'Public Administration' sector [10]. The inequalities in the country are particularly relevant when it comes to access to basic sanitation, sewage treatment and potable water [63]. The provision of services on items such as water and electricity still follow a poorly effective and highly centralized approach. According to Lepre and Castillo [35], in the Northeast region, one of the poorest in the country, many communities still live in the dark and distant from sources of drinking water. Whilst Brazil is one of the world's leading producers of hydroelectric power, with a current capacity of about 260,000 megawatts, the most relevant initiatives in the energy sector are those directed towards large-scale facilities [63].

In order to reverse this situation, there is a growing number of community-based initiatives, start-ups and NGOs that are investing in more decentralized or distributed approaches, deploying and implementing water and energy solutions with small and flexible localized units. New regulation is stimulating the construction of small-scale hydroelectric plants, which in Brazil are defined as those with a capacity of 5 to 30 MW and an area of reservoir limited to 13 km^2. From 331 small-scale plants in 1999, the country reached 1129 in 2019, according to ABRAPCH [1].

Industry in Brazil follows a Distributed Economy in those sectors with lower demands on technology or with lower demands on capital investment, enabling individuals or small companies to start their own business. This is the case in the clothing and textile sector, for instance. Brazil is a country where all stages of the clothing supply chain can be found within the country borders, from fibre production to semi-processed products (yarn and fabrics with their finishing processes) and final products. Industrial clusters in this sector are good examples of decentralized or distributed approaches to the economy. These clusters are composed of a variety of company sizes and types, including cooperatives and/or craftworkers, organized in close proximity to customers and suppliers, contributing to optimize their production and logistic processes.

In contrast, in the agricultural sector, there is a mix of centralized, decentralized and distributed approaches, operating simultaneously across the supply chain. Part of the expansion of the agribusiness sector occurred at the expense of the environment, including the Amazon. It is quite common that investment in this sector prioritizes large-scale farms, huge silos that often stock grains for more than a year waiting for better international prices, and large ports with correspondent large ships to transport commodities across the oceans. However, in this same agricultural sector there are federal, state and municipal initiatives directed towards family agriculture, which is highly distributed in its essence, with around 800 thousand rural inhabitants being assisted with credit, research and extension programmes [42]. These small-scale local farmers supply food to rural communities, schools and on urban street markets, in direct contact with their final consumers.

8.2   A Chinese Perspective

In China, sustainable development has become a social consensus. Meanwhile, the relevant concepts of sustainable development have been widely recognized at all levels of society, and these concepts are consistent with the principles of the DE to a certain extent. On the other hand, China can benefit from its development in the Internet field, and the promotion and implementation of a distributed economy are possible. We can see that technological advances are rapidly affecting and changing China's consumption patterns. Manufacturing, energy, water, food and information/knowledge production industries are showing decentralized/distributed trends and potentials to varying degrees and will bring challenges to the mainstream economic model. However, it should also be noted that China's current development success has actually relied on a central development model. Therefore, for a long time to come, in China, the status of this central economic development model will remain unshakable. All stakeholders committed to promoting China's sustainable transformation need to think carefully and rationally about the role of the distributed economy.

We also need to acknowledge that the sustainable development of various regions in China is not balanced, and there is a clear difference in sustainable development between second/third-tier cities and first-tier cities (Beijing, Shanghai, Guangzhou and Shenzhen). Especially in terms of sustainable production and consumption, although China has been actively promoting cleaner production and green consumption lifestyles, China's economic development mode is still in a relatively extensive stage. Consumption and high pollutant emissions still exist. On the other hand, the public's awareness of green consumption and production needs to be further improved. From another perspective, this can also be seen as an excellent opportunity for a distributed economy to realize its sustainable potential. As a large and dynamic country, China is likely to have extensive and in-depth development and actions in many areas of the distributed economy in the future [68].

8.3   A Finnish Perspective

In Finland, certain concepts related to a more sustainable society have become prominent, which are grounded on principles that are compatible with those of Distributed Economies. This is not surprising, as DE was developed in the neighbouring country of Sweden, and much of northern Europe has experienced the negative economic effects of manufacturing that has moved offshore to regions with cheaper labour and raw materials while recognizing that our consumption patterns are also outsourcing pollution and bad working conditions to these regions. In Finland, this was especially visible in the fashion and textiles sector. DE principles related to revitalizing the economy, regional collaboration on high-value-added, high-quality products using local raw materials and resources (knowledge, manufacturing capabilities and skills), are therefore easily applied. The most popular economic revitalization concept that robustly embeds sustainability considerations in Finland is that of a Circular Economy (or Circular Bio-Economy). In this vision, local resources related to biomass circulate as biological nutrients in the organic cycle of the economy, adding value where possible through upcycling and cascading. Stakeholders, companies, research institutes, investors and customers, collaborate in production, research and innovation, in order to diversify the Finnish economy and strengthen its resilience. Therefore, Finland as a region with a particular industrial history would find many aspects of Distributed Economies strategically attractive.

8.4 An Indian Perspective

Pre-colonial industrialization in India was largely based on distributed, village-based economies, even for global trade in manufactured goods like textiles and handicrafts. However, colonization and the subsequent post-colonial industrialization created a push towards centralized global and monopolistic manufacturing systems which denuded the network of local production economies. Over the last few decades, there has been cross-sectoral movement back towards distributed economies motivated mainly by issues of livelihood generation and economic empowerment by tapping into urban markets to develop opportunities for rural economies.

Distributed production systems were revived on a large scale through cooperative dairy companies like Amul and traditional food companies like Lijjat Papad, formed in the late 1940s and early 1950s, which have managed to develop vast networks of village-based production units. These companies set the template for distributed economies which, over the past two decades, have developed in diverse sectors like fashion and textiles, handicrafts, food processing, energy production and water management among others, resulting in tens of thousands of people being financially empowered and in a shift towards more environmentally and socio-ethically conscious consumption patterns, as well as a growing interest in traditional and indigenous aesthetics and lifestyles.

In urban India, distributed economies have been powered by technological aggregator platforms mainly in the service sector in industries ranging across design and architecture, construction and maintenance, transportation, food and beverages and hospitality. Environmentally sound practices are increasingly being incorporated into these platforms.

While these developments are varied and exciting, their theorization within the discourse of distributed economies remains at a nascent stage. The challenge will be to understand how these economies function in relation to each other and how they can work within larger economic and ecological systems.

8.5 A Mexican Perspective

We can distinguish three important factors in the Mexican economy:

1. Large investments are made by transnational industries that are concentrated in specific states. According to INEGI (National System of Statistical and Geographical Information in Mexico), the manufacturing industry has made the largest contribution to state GDP in Coahuila de Zaragoza, Querétaro, State of Mexico, Aguascalientes, Guanajuato, Puebla and San Luis Potosí, which coincides with the investment plans reported by a survey published by Manufactura MX [37].
2. The traditional production models that have been able to resist Mexico's incursion into global markets are those oriented towards a Distributed Economy.

3. The informal sector takes an important role, both because of its scale and because
    it mainly focuses on the satisfaction of local markets, one of the key characteristics
    of Distributed Economy models.

In Mexico, the industrialization process has focused on development poles in
specific geographical areas, which has created impoverished regions where economic
activities develop with difficulty. The industrialization process in Mexico has not
always been the result of an international state policy; sometimes it has responded
to industry push and the changing conditions of the environment [25]. On the other
hand, after the entry into force of NAFTA (North American Free Trade Agreement),
industrialization has been driven by the creation of global supply chains, where the
strategy focuses on opening up to foreign trade [25] and not to the satisfaction of
local markets.

The investment plans of the manufacturing sector are settled in eight states
(national regions). It is not yet a priority to enter the three special economic zones
(EEZs) declared in 2015 by the federal government to boost development in regions
with greater social and economic lags in the country, according to the study [37]. This
indicates that although there are public policy efforts to generate development poles
that move closer to the decentralized model, the investment plans of the companies
are oriented towards maintaining a traditional industrialization model. The survey
applied to 812 Mexican business leaders nationwide, of large and medium-sized
companies from various industries, established in the country, reveals that 55% of
those interviewed are taking their company to a state in Mexico. In 2015, that estimate
was 63 percent [37].

However, it is possible to find cases of models closer to distributed economies
that respond to the satisfaction of local markets. Nevertheless, they are currently in
danger because of public policy trends aimed at impacting global markets. To take
one example, the Colonia Morelos neighbourhood in Mexico City is so large that it
contains two important boroughs: the Cuauhtémoc and the Venustiano Carranza. It
is currently one of the most important areas for drug trafficking, which has made it a
violent area; however, its commercial activities dating from the last century (1881)
still prevail. At that time, its inhabitants were engaged in the manufacture of shoes,
a trade of great tradition and which continues in one of its neighbourhoods, Tepito.
Currently, along the principal avenue of that zone, several supply stores related with
the manufacture of shoes and bags are established, as well as workshops that offer
Product-Service Systems i.e. manufacturing parts of the shoe production process are
offered. In other words, shoemakers who do not have sewing machinery, for example,
can send their pre-cut pieces to local workshops, which offer sewing services. In this
way, finished products are offered in the local Granaditas Market.

In Mexico, local markets are served not only by the formal sector: 76 out of
every 100 pesos generated from GDP are produced by 42% of all formal jobs and 24
pesos are generated by 58% of informal jobs. Informality in Mexico is widespread
and, in particular, much more widespread than in other countries in the region. High
informality is worrying because it denotes an inadequate distribution of resources (in
particular labour) and an extremely inefficient use of government services, which can

compromise the country's growth prospects. Mexico's principal challenge would be focused on finding an efficient strategy to turn back to local markets through DE.

References

1. ABRAPCH (Associação Brasileira de PCHs e CGHs) (2019) Número de PCHs em operação no Brasil. https://abrapch.org.br/faq/numero-pchs-em-operacao-no-brasil/. Accessed 25 Nov 2019
2. Alanne K, Saari A (2006) Distributed energy generation and sustainable development. Renew Sustain Energy Rev 10(6):539–558
3. Andrews G (2008) The slow food story: politics and pleasure. Pluto Press, London
4. Angeles-Martinez L, Theodoropoulos C, Lopez-Quiroga E, Fryer P, Bakalis S (2018) The honeycomb model: A platform for systematic analysis of different manufacturing scenarios for fast-moving consumer goods. J Clean Prod 193:315–326
5. Beltagui A, Kunz N, Gold S (2020) The role of 3D printing and open design on adoption of socially sustainable supply chain innovation. Int J Prod Econ 221:107462
6. Benkler Y (2006) The wealth of networks: how social production transforms markets and freedom. Yale University Press, New Haven, CA
7. Bessière D, Charnley F, Tiwari A, Moreno MA (2019) A vision of re-distributed manufacturing for the UK's consumer goods industry. Prod Plann Control 30(7):555–567
8. Bogers M, Hadar R, Bilberg A (2016) Additive manufacturing for consumer-centric business models: Implications for supply chains in consumer goods manufacturing. Technol Forecast Soc Chang 102:225–239
9. Brede M, de Vries BJM (2009) Networks that optimize a trade-off between efficiency and dynamical resilience. Phys Lett A 373(43):3910–3914
10. Cardoso V, Perobelli F (2014) Evaluation of the structure of the services sector in Brazil: a regional approach, ERSA conference papers ersa14 p 865, European Regional Science Association
11. Carson KA (2010) The Homebrew industrial revolution: a low-overhead manifesto. Booksurge
12. Castells M (2000) The rise of the network society, 2nd edn. Wiley-Blackwell, Malden, MA
13. Ceschin A (2010) A natureza como limite da economia: a contribuição de Nicholas Georgescu-Roegen. Editora Senac São Paulo/Edusp, São Paulo
14. Chaurey A, Krithika PR, Palit D, Rakesh S, Sovacool BK (2012) New partnerships and business models for facilitating energy access. Energy Policy 47:48–55
15. Costa C, Pelegrini A (2018) Design Distribuído: novas práticas e competências para o design pós-industrial. In: Anais do 13° P&D Design. Univille, Joinville
16. Crul M, Diehl JC (2006) Design for sustainability (D4S): a step-by-step approach. Modules, United Nations Environment Programme (UNEP)
17. Daly H (2010) Crescimento se tornou antieconômico. Revista Época Negócios, julho. http://epocanegocios.globo.com/Revista/Common/0,EMI151601-16381-1,00-CRESCIMENTO+SE+TORNOU+ANTIECONOMICO+DIZ+HERMAN+DALY+PAI+DA+ECONOMIA+ECOLOGI.html+PAI +DA+ECONOMIA+ECOLOGI.html. Accessed 27 May 2020
18. De Vere I (2013) Industrial design 2.0: a renaissance. In: International conference on engineering and product design education 5 & 6, Dublin Institute of Technology, Dublin
19. Dickie IB (2018) Proposition of a reference model of Crowd-design for sustainability. Post Graduation Program in Design [Doctoral Dissertation]. Federal University of Parana, Curitiba, Brazil
20. Dreyer B, Lüdeke-Freund F, Hamann R, Faccer K (2017) Upsides and downsides of the sharing economy: Collaborative consumption business models' stakeholder value impacts and their relationship to context. Technol Forecast Soc Chang 125:87–104

21. Emili S, Ceschin F, Harrison D (2016) Product-Service Systems applied to Distributed Renewable Energy: a classification system and 15 archetypal models. Energy Sustain Dev 32:71–98
22. Emili S (2017) Designing product-service systems applied to distributed renewable energy in low-income and developing contexts: a strategic design toolkit. PhD Thesis, Brunel University London
23. Escobar A (2018) Designs for the Pluriverse: radical interdependence, autonomy, and the making of worlds. Duke University Press, Durham, US
24. Freeman R, McMahon C, Godfrey P (2017) An exploration of the potential for re-distributed manufacturing to contribute to a sustainable, resilient city. Int J Sustain Eng 10(4–5):260–271
25. Flores Kelly J (2018) México piensa+. Ediciones Felou, Mexico City
26. Folke C (2016) Resilience (Republished). Ecol Soc 21(4):Article 44
27. Gaiger LI (2008) A dimensão empreendedora da economia solidária: Notas para um debate necessário. Otra Economia 11(3)
28. Hawken P, Lovins A, Lovins LH (1999) Natural capitalism: creating the next industrial revolution. Little, Brown, and Company, Boston
29. Illich I (1973) Tools for conviviality. Harper and Row, New York
30. IIIEE (International Institute for Industrial Environmental Economics) (2009) The future is distributed: a vision of sustainable economies. IIIEE, Lund
31. Johansson A, Kisch P, Mirata M (2005) Distributed economies – A new engine for innovation. J Clean Prod 13(10–11):971–979
32. Kaneko K, Kishita Y, Umeda Y (2018) Toward developing a design method of personalization: Proposal of a personalization procedure. Procedia CIRP 69:740–745
33. Kaundinya DP, Balachandra P, Ravindranath NH (2009) Grid-connected versus stand-alone energy systems for decentralized power—A review of literature. Renew Sustain Energy Rev 13(8):2041–2050
34. Kumar M, Tsolakis N, Agarwal A, Srai JS (2020) Developing distributed manufacturing strategies from the perspective of a product-process matrix. Int J Prod Econ 219:1–17
35. Lepre P, Castillo L (2019) The third sector as a vector to foster distributed design and distributed economy initiatives: a case study of inclusive, ethical and sustainable social development in emerging economies. In: Ambrosio M, Vezzoli C (eds) Designing sustainability for all - 3rd LeNS world distributed conference proceedings. Milano, IT, Edizioni Poli.design
36. Mandelli S, Mereu R (2014) Distributed generation for access to electricity: "Off-main-grid" systems from home-based to microgrid. In: Colombo E et al (eds) Renewable energy for unleashing sustainable development. Springer, Heidelberg, pp 75–97
37. Manufactura MX (2016) 8 Estados más atractivos para los industriales. https://manufactura.mx/industria/2016/04/26/los-8-estados-mas-atractivos-para-la-industria. Accessed 25 Nov 2019
38. Martin M, Lazarevic D, Gullström C (2019) Assessing the environmental potential of collaborative consumption: peer-to-peer product sharing in Hammarby Sjöstad, Sweden. Sustainability 11(1)
39. Mathai MV (2012) Green economy and growth: Fiddling while Rome Burns? United Nations University. http://unu.edu/articles/science-technologysociety/green-economy-and-growth-fiddling-while-rome-burns. Accessed 27 Jan 2012
40. McKee D (2017) Neoliberalism and the legality of peer platform markets. Environ Innov Soc Transit 23:105–113
41. Mirata M, Nilsson H, Kuisma J (2005) Production systems aligned with distributed economies: examples from energy and biomass sectors. J Clean Prod 13(10–11):981–991
42. MTE (Ministério do Trabalho e Emprego) (2012) O que é Economia Solidária. http://www.mte.gov.br/ecosolidaria/ecosolidaria_oque.asp. Accessed 14 Oct 2012
43. Mumford L (1934) Technics and civilization. Harcourt, Brace, and Company, New York
44. Murillo D, Buckland H, Val E (2017) When the sharing economy becomes neoliberalism on steroids: unravelling the controversies. Technol Forecast Soc Chang 125:66–76
45. Napier E, Sanguineti F (2018) Fashion Merchandisers' Slash and Burn dilemma: a consequence of over production and excessive waste? Rutgers Bus Rev 3(2):159–174

46. Niinimäki K, Peters G, Dahlbo H, Perry P, Rissanen T, Gwilt A (2020) The environmental price of fast fashion. Nat Rev Earth Environ 1(4):189–200
47. Pearson H, Noble G, Hawkins J (2013) Re-distributed manufacturing workshop report. EPSRC, UK
48. Peças Lopes JA, Hatziargyriou N, Mutale J, Djapic P, Jenkins N (2007) Integrating distributed generation into electric power systems: a review of drivers, challenges and opportunities. Electr Power Syst Res 77(9):1189–1203
49. Petrulaityte A, Ceschin F, Pei E, Harrison D (2017) Supporting sustainable product-service system implementation through distributed manufacturing. In: The 9th CIRP industrial product-service system conference: circular perspectives on product/service-systems, Denmark
50. Pizzol M (2015) Life cycle assessment and the resilience of product systems. J Ind Ecol 19(2):296–306
51. Rauch E, Dallasega P, Matt DT (2016) Sustainable production in emerging markets through Distributed Manufacturing Systems (DMS). J Clean Prod 135:127–138
52. Rolland S (2011) Rural electrification with renewable energy: technologies, quality standards and business models. Alliance for Rural Electrification, Brussels, Belgium
53. Botsman R, Rogers R (2011) What's mine is yours: the rise of collaborative consumption. Collins, London
54. Rowe PCM (2017) Beyond Uber and Airbnb: the social economy of collaborative consumption. Social Media Soc 3(2):2056305117706784
55. Sachs I (2002) Caminhos para o desenvolvimento sustentável. Garamond, Rio de Janeiro
56. dos Santos A (2018) Theoretical foundations on SPSS and DE: a survey among members of the learning network on sustainability (LeNS). Report on Round 02 of Questioning. Learning Network on Sustainability, Curitiba
57. dos Santos A, Braga AE, Sampaio CP, Andrade ER, Merino EAD, Trein F, Duarte GG, Rosa IM, Massaro JG, Lepre PR, Noronha R, Engler R, Vasques RA, Nunes VGA (2019) Design para a Sustentabilidade: Dimensão Econômica, 1st edn. Editora Insight. v, Curitiba, p 1
58. Schillebeeckx SJD, Parikh P, Bansal R, George G (2012) An integrated framework for rural electrification: Adopting a user-centric approach to business model development. Energy Policy 48:687–697
59. Scholz T, Schneider N (eds) (2016) Ours to hack and to own: the rise of platform cooperativism, a new vision for the future of work and a fairer internet. OR Books, New York
60. Sen A (1999) Sobre ética e economia. Companhia das Letras, São Paulo
61. Srai JS, Kumar M, Graham G, Phillips W, Tooze J, Tiwari A, Ford S, Beecher P, Raj B, Gregory M, Tiwari M, Ravi B, Neely A, Shankar R (2016) Distributed manufacturing: Scope, challenges and opportunities. Int J Prod Res 54(23):6917–6935
62. Stahel WR (1986) Hidden innovation: R&D in a sustainable society. Science & Public Policy 13(4):196–203
63. Tundisi JG (2008) Water resources in the future: problems and solutions. Estudos Avançados 22(63) São Paulo
64. da Veiga, J.E. (2005). O Prelúdio do Desenvolvimento Sustentável. In: CAVC, Economia Brasileira: Perspectivas do Desenvolvimento, pp. 243–266. http://www.zeeli.pro.br/Livros/2005_b_preludio_%20desenvolvimento_sustentavel.pdf. Accessed 27 May 2020
65. Vezzoli C, Ceschin F, Osanjo L, M'Rithaa MK, Moalosi R, Nakazibwe V, Diehl JC (2018) Designing sustainable energy for all: sustainable product-service system design applied to distributed renewable energy. Springer, London
66. Vezzoli C (2010) System design for sustainability: a promising approach for low and middle-income contexts. In: International forum Design for Sustainability in Emerging Economies. June 2–5, Autonomous Metropolitan University of Mexico City
67. Vezzoli C, Delfino E, Amollo Ambole L (2014) System design for sustainable energy for all. A new challenging role for design to foster sustainable development http://dx.doi.org/10.7577/formakademisk.791
68. Xia N, Liu X (2016) The new context of design: research on the sustainability of distributed economy. Zhuang Shi (12)

Integrating S.PSS and DE

Ranjani Balasubramanian, Carlo Vezzoli, Fabrizio Ceschin, Jacob Matthew, Abhijit Sinha, and Christoph Neusiedl

1 Introduction to S.PSS Applied to DE: Sustainable Opportunities

The combination of Sustainable Product-Service Systems (S.PSS) and Distributed Economies (DE) has been considered as a promising mode of developing sustainability through regional resilience and by empowering a shift to a more localized economic model [11]. Especially in regions with significant middle- and low-income populations, DE provides possibilities for increased localized employment generation, and many such schemes have been implemented in both urban and rural areas. In underserved regions, this could help decrease emigration and develop better services in these economies. The LeNSin project studied the shift to S.PSS as a mode of DE designing and delivering. The win-win sustainability benefits could be summarized as follows.

S.PSS is a promising approach to diffuse DE in low/middle-income (all) contexts, because it reduces/cuts both the initial (capital) cost of DE product/equipment purchasing (that may be unaffordable) and the running cost for maintenance, repair, upgrade, etc. of such DE hardware (that may cause

R. Balasubramanian (✉)
Srishti Institute of Art, Design and Technology, Bengaluru, India
e-mail: ranjani.b@srishti.ac.in; ranjani.bms@gmail.com

C. Vezzoli
Design Department, Politecnico di Milano, Milan, Italy

F. Ceschin
Department of Design, Brunel University London, London, UK

J. Matthew
Industree Foundation, Bengaluru, India

A. Sinha · C. Neusiedl
Project DEFY, Bengaluru, India

C. Vezzoli et al. (eds.), *Designing Sustainability for All*, Lecture Notes in Mechanical Engineering, https://doi.org/10.1007/978-3-030-66300-1_3

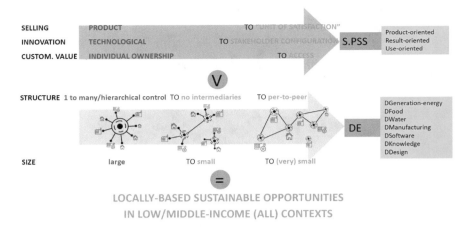

Fig. 1 The coupling of the two win-win sustainability paradigm shifts represented by S.PSS and DE

the interruption of use) increasing local employment and related skills. Furthermore, by offering a DE system adopting an S.PSS model, the producer/provider is economically incentivized to design low environmentally impacting DE products/equipment. Finally, S.PSS applied to DE is a promising key leverage for a sustainable development process for all aiming at democratizing access to resources, goods and services.

These win-win potentials are based, in fact, on the coupling of the two win-win sustainability paradigm shifts of S.PSS and DE we have already discussed in the previous chapters of S.PSS and DE we have already discussed in the previous chapters (see Fig. 1):

1. The shift from a traditional product sale model to S.PSS, i.e. the shift of customer perceived value from individual ownership to access to a mix of products and services (systems) fulfilling a given unit of satisfaction;
2. The shift from centralized to decentralized/distributed systems in which a small-scale unit of production is locally based, i.e. nearby or at the point of use, and where the user can become a producer.

Further consideration could be made in relation to the increased access to the internet and digital infrastuctures and tools combined with the projected development of distributed technologies, such as 3D printing, which significantly increase the potential and ease of setting up these Distributed Economies. In areas with low income, even basic internet penetration has opened up possibilities to access knowledge and know-how to set up distributed networks. A number of organizations and governments are supporting the set-up of such networks in low- and middle-income regions, and the main focus is to develop affordable systems with the aid of technology that requires lower investment cost. In middle- to higher-income regions, the likelihood of using more capital-intensive processes (like 3D printing manufacturing) is higher and there is a push to develop DE networks with the aid of technology. However, it seems that technology and access to information sharing systems will be key to developing scalable and replicable DE. With the accelerated pace of technological penetration, it is possible to envision what an S.PSS would look like applied in a DE format.

Sustainable value-adding PSS can only be created taking into account every life cycle stage of products and services [8]. Distributed Manufacturing, for example, applied to *near-future* scenarios addresses each S.PSS life cycle stage, thus showing the potential to improve PSS development from the sustainability point of view:

- The *design* stage predominantly benefits from collaboration between PSS provider and customer, enabled by connectivity through digital channels and physical interaction in local production facilities, which results in better S.PSS acceptance.
- The *material production (pre-production)* and *production* stages benefit from the distribution of manufacturing facilities, equipped with digitally connected manufacturing technology. The ability to send digital production files to remote locations, for example, allows PSS companies to produce products and spare parts in close proximity to customers and/or resources, thus reducing the environmental impact of distribution.
- The *use* stage is supported with the largest number of near-future scenarios tackling on-site and on-time provision of maintenance services and empowering customers to maintain, repair, update, upgrade and re-manufacture the products included in the S.PSS solution.
- The *end-of-life* phase is facilitated by the application of sensor technology, which helps to indicate products' and components' end-of-life by alerting PSS providers and customers. Finally, a distributed network of localized recycling facilities eases product collection, recycling and/or energy recovery.

This chapter examines case studies of S.PSS applied to DE (both Distributed and Decentralized production units) from across the globe. It is important to note nevertheless that it is challenging to clearly define and categorize the case studies, as most of them consist of varying degrees of PSS or DE with different types of interactions. These could, however, be used to develop a categorical understanding of S.PSS applied to DE.

2 Case Studies of S.PSS and DE Integration

2.1 S.PSS and Distributed Energy Generation (DG)

As discussed in Chap. 2, Distributed energy Generation (DG) represents a promising strategy to provide energy access with a range of sustainability benefits. However promising, the implementation of DG solutions should not only focus on the technical

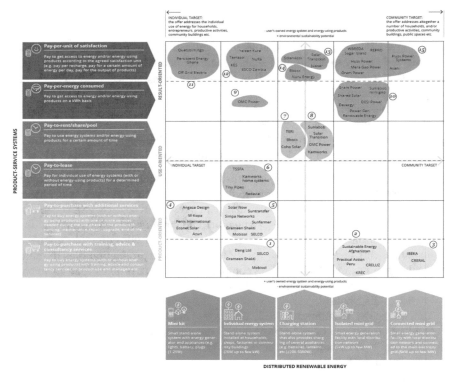

Fig. 2 Archetypal models of PSS applied to DG in low- and middle-income contexts [2, 3]

aspects. There are other aspects which are crucial for the success of DG solutions. Most of the unsuccessful cases of DG are linked to problems such as the lack of a maintenance and repair network, lack of understanding of user needs or lack of a proper business model [3, 10]. For this reason, an S.PSS system design approach should be adopted. This means that, in addition to energy technology, the stakeholder value chain, the product-service combination and the business model aspects should be taken into consideration and integrated into systemic solutions [6, 9]. S.PSS applied to DG can be categorized in 15 archetypal models (Fig. 2) [2, 3]:

1. Selling individual energy systems with advice and training services;
2. Offering advice and training services for community-owned and-managed isolated mini-grids;
3. Offering advice and training services for community-owned and-managed connected mini-grids;
4. Selling mini-kits with additional services;
5. Selling individual energy systems with additional services;
6. Offering individual energy systems (and energy-consuming products) in leasing;
7. Renting energy-using products through entrepreneur-owned and-managed charging stations;

8. Renting energy-using products through entrepreneur- or community-managed charging stations;
9. Offering access to energy (and energy-using products) on a pay-per-consumption basis through individual energy systems;
10. Offering access to energy (and energy-using products) on a pay-per-consumption basis through isolated mini-grids;
11. Offering access to energy & energy-using products on a pay-per-unit of satisfaction basis through mini kits;
12. Offering access to energy (and energy-using products) on a pay-per-unit of satisfaction basis through individual energy systems;
13. Offering access to energy-using products through community- or entrepreneur-managed charging stations on a pay-per-unit of satisfaction basis;
14. Offering recharging services through entrepreneur-owned & -managed charging stations;
15. Offering access to energy (and energy-using products) on a pay-per-unit of satisfaction basis through mini-grids.

Several case studies of S.PSS and Distributed energy Generation (DG) are presented below.

SELCO (example of archetypal model 1 and 5)
  Active since: 1995
  Provider/s: SELCO and local community agents
  Customers: Rural Households/Communities, Institutions
  S.PSS Type: Product-oriented S.PSS
  DE configuration: Distributed and Decentralized energy Generation
  Products: Solar Home Lighting, Solar Water Heater, Solar Inverter Systems, DC Home Appliances like Butter Churners, Grinders, etc.
  Services: Product customization, installation, maintenance and repair, community training, tailoring financing options, advisory and capacity building.
  Payments: Pay for product-service-system
  Resource: Solar Energy
  Location: India

SELCO is a rural energy service social enterprise that provides affordable and environment-friendly energy services to rural households. SELCO produces solar Product-Service Systems for individuals, communities or institutions. The ultimate aim of the company is to provide affordable rural electrification through renewable sources and to achieve this, SELCO provides services that include financing consultancy, customized product configurations, training, maintenance and repair.

The company also creates additional distributed economies by training local youth for maintenance of the systems, supporting local entrepreneurs who can buy the

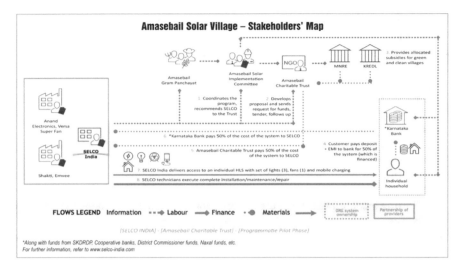

Fig. 3   System map of a SELCO S.PSS for a village in India. (*Source:* renewablewatch.in)

system and develop a livelihood by providing charging services and connecting them to financial services. The company has also diversified into producing solar energy-powered products like sewing machines and photocopy machines which can further develop into opportunities for distributed manufacturing.

The most striking characteristic of the company is its targeted user group and diversified Product-Service System in a standard distributed format, involving multiple relevant stakeholders (Figs. 3 and 4).

Solarkiosk (example of archetypal model 15)

Active since: 2011

Provider/s: Solarkiosk Solutions GmbH (E-HUBB and related equipment); local subsidiary (installation, maintenance and repair)

Customers: Solarkiosk local subsidiaries (own model), international organizations (B2B)

S.PSS Type: Use-oriented (B2B), Result-oriented (B2C)

DE configuration: Decentralized energy Generation

Products: E-HUBB, Solar Pico systems, Solar Home Systems, PAYG systems, other products

Services: Project based design, production, installation, maintenance, engineering

Payments: E-HUBB is in ownership of Solarkiosk (own model), Project budgets (B2B sales)

Fig. 4 Solar powered cow milking unit. (*Source* SELCO)

Payments: E-Hubb is given for free (B2B), Pay per use (B2C)

Resource: Solar Energy

Location: Ethiopia, Kenya, Rwanda, Tanzania, projects realized in 11 other countries

The company targets local entrepreneurs, especially women, for the provision of energy services through charging stations. Solarkiosk designs and installs the E-Hubb, a charging station provided with solar panels and energy-consuming products and recruits a local entrepreneur who manages the system and appliances. Due to the modular configuration of the station, he/she can provide a wide range of energy-dependent services such as internet connectivity, water purification, copying, printing and scanning. Customers pay for the service they need: pay to print, pay to get purified water, pay for internet access and other services (Figs. 5 and 6).

2.2   S.PSS and Distributed Food Production (DF)

If small-scale Food production units (DF) are connected with other DF (e.g. to share food overproduction), they become a Distributed Food Production Network (DFN), which may in turn be connected with nearby similar networks. If properly designed they hold promise to promote locally based sustainability, i.e. Sustainable Distributed Food production (S.DF) systems.

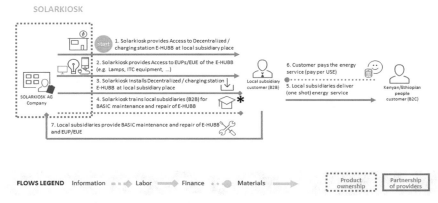

Fig. 5 Solarkiosk system map

Fig. 6 Solarkiosk Ethiopia, 2011

The new configuration results in reducing the need for transportation of food from outside the city. It also minimizes the use of packaging and storage. Producers and consumers connect with each other, and consumers assume a new role as co-producers who have the opportunity to learn more about local food production, while they get involved in the process of growing food and raising animals. In this way, urban farming empowers communities to share knowledge and diversity, keeping alive food traditions and local food heritage.

Several case studies of S.PSS and Distributed Food production (DF) are presented below.

PickYourOwn

Active since: 2008

Provider/s: Farmers

Customers: Home users (B2C), Commercial Business (B2B)

S.PSS Type: Use-oriented S.PSS

DE configuration: Decentralized Food production

Products: Fruits and vegetables

Services: Use of kitchen and canneries facilities, channel for collaboration, education, consultancy and certified production.

Payments: Pay per period/time or pay per produced unit or each process for the use of kitchen/caning facilities. Pay per product (farm)

Location: USA

Pick-your-own is an idea for home or commercial users to pick their own fruit from the local farms near them and use them in distributed food production. The website Pick-YourOwn.org lists farms located all around the country who provide their products to be sold with the pick-your-own concept. On the website there is also a calendar of the harvesting time of different products. The home users or commercial users can pick fresh vegetables and fruits on these farms and produce canned/bottled/packed products using the kitchen/canning facilities that are in shared/community/commercial kitchens and canneries. The users can produce products for their own use as well as to sell or exchange. While most of the facilities are more oriented towards home users, some are oriented towards commercial users. In most sites, they also provide information and education for production in their facilities. Some have licenses that enable users to produce for commercial use. They also function as a hub for users to meet, collaborate and learn from each other. The two common payment methods are pay per period, pay per produced unit/each process or a combination of both. In this case, while the production and consumption of vegetables is distributed, the production also includes the service of fresh food combined with the customer experience of handpicking it. It also reduces the need for packaging and transportation for the producer as well as ensuring a fair price for the produce (Figs. 7 and 8).

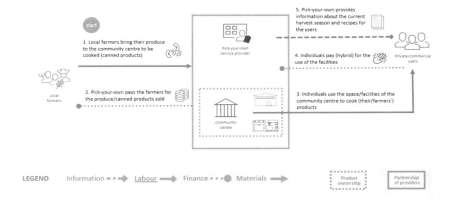

Fig. 7   Pick-your-own system map

Fig. 8   Example of a Pick Your Own farm, 2008

FoodyBuddy

Active since: 2017

Provider/s: Home Chefs

Customers: Hungry Individual Consumers

S.PSS Type: Use-oriented S.PSS

DE configuration: Decentralized Food production

Products: Fresh Cooked Meals

Services: Aggregator platform connecting home-based chefs to customers

Payments: Pay per product delivered

Location: India

Foodybuddy is a neighbour food network that connects home-based chefs to customers at a hyper-local level. The app allows home chefs to decide upon the menu, number of portions, days of sale, timings and pricing of meals, allowing for flexible income generation. The consumer has the advantage of viewing a daily menu of food on offer in their neighbourhood or apartment complex and communicate with the seller on the app.

Since this system works at a hyper-local level, it eliminates the need for transportation. The food is either delivered by the seller or picked up by the consumer. This also allows the seller and the customer to interact personally and develop connections within the neighbourhood. There is an opportunity to connect this service with existing delivery services if it is so required, as an example of a networked distributed system (Fig. 9).

Fig. 9 Service onboarding on the Foodybuddy App. (*Source* Foodybuddy App)

2.3   S.PSS and Distributed Water Management (DW)

Water management is an area that is increasingly witnessing the dangers of the failure of excessive centralization. With the development of regional scale systems of water management like large dams and reservoirs—primarily for agriculture and power generation—there is evidence of increasing negative impacts on ecosystems [1].

Decentralized solutions for water collection, storage, treatment and use are being revived from traditional systems or developed as new solutions to cater to the needs of vast populations underserved by centralized water management projects. In many parts of the world, limited access to fresh water is also becoming an issue of political contention which disenfranchises vast numbers of people from the process of water management and access. In this scenario, provision of clean water as a service has great potential for developing distributed models of management and access that also empowers communities to be self-sufficient and fosters community-based income generation models. There are organizations that work with community-based catchment management, water storage and treatment.

Several examples of organizations that provide potable water to underserved communities in an S.PSS and Distributed Water management (DW) are presented below.

Piramal Sarvajal

Provider/s: Piramal Sarvajal with local franchisees

Customers: Underserved rural and urban communities

S.PSS Type: Use-oriented S.PSS

DE configuration: Decentralized Water Management Network

Products: Water ATMs, Water purifiers, Water Quality Monitoring Units

Services: Community awareness and training, centralized water quality monitoring, water delivery system

Payments: Pay per use

Location: India

Piramal Sarvajal sets up community-level solutions that are locally operated but centrally managed on a market-based pay-per-use system. The last-mile operational accountability is ensured by developing and deploying remotely monitored and controlled drinking water purification systems. Piramal Sarvajal's other product is the Water ATM: a solar-powered, cloud-connected, smart card-based automatic water vending machine.

While the water purification and delivery systems follow a pay-per-use S.PSS model with an emphasis on socio-ethical and economic sustainability, the distributed system of water purification also allows for developing distributed economies through community-based franchisees (Fig. 10).

Ecosoftt and Gram Vikas

Provider/s: Ecosoftt and Gram Vikas (Partner NGO)

Customers: People from villages without access to clean water

S.PSS Type: Use-oriented S.PSS

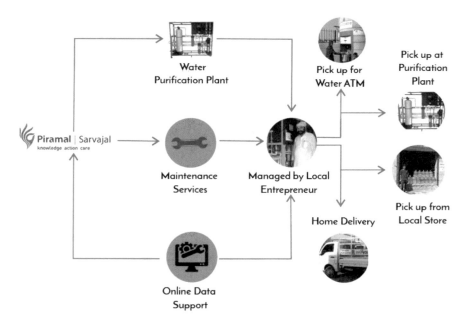

Fig. 10 System map of Piramal Sarvajal. (*Source* Piramal Sarvajal)

Fig. 11 Decentralized clean
water systems for villages

DE configuration: Decentralized Water Management

Products: Equipment to build infrastructure

Services: Access to drinking water, toilets, bathing rooms and wastewater management systems

Payments: Pay per use

Location: India

Ecosoft in collaboration with Gram Vikas (an NGO) provides equipment + training
for local users to build infrastructure to take water from underground and provide
decentralized access to clean water in the village. The package consists of access to
drinking water, toilets, bathing rooms and a wastewater management system. There
is no investment cost for the local community; they pay to Ecosoft according to the
amount of water they consume. The package also includes training for maintenance
and providing equipment in case of replacement needed (Fig. 11).

2.4 S.PSS and Distributed Manufacturing (DM)

Current manufacturing and supply chains have become extremely efficient global
systems that draw labour, material, production and assembly from centres around
the world. These supply chains have been honed to function at maximum efficiency.
However, it is also notable that this efficiency comes at the cost of redundancy and

resilience. Global events like pandemics are proving that there is a dire need for developing more resilient and localized systems of production and distribution.

As mentioned in Chap. 2, three key features of digital manufacturing have been identified as:

- Localization of manufacturing units;
- Application of physical and digital technologies;
- Customer orientation.

Distributed manufacturing allows more people to develop local livelihood opportunities that can contribute towards building economic sustainability. A movement away from extractive global manufacturing processes marks a potential to develop ecological and socio-ethical sustainability in local communities.

Several case studies of S.PSS and Distributed Manufacturing (DM) are presented below.

StrataSys Leasing

Active since: 2011

Provider/s: StrataSys

Customers: Small and large enterprises, makers, designers, engineers (B2B)

S.PSS Type: Use-oriented and Result-Oriented S.PSS

DE configuration: Decentralized Manufactoring

Products: 3D printers, start-up supplies, support removal system, cleaning agent

Services: Optional services (system operation, inhouse support, education, project implementation, consulting)

Payments: Pay per period (fixed cost)

Location: USA and Israel (headquarters), Canada, Brazil, Mexico, Germany, Japan, Korea, China, Singapore, India

StrataSys manufactures 3D printers and offers 3D production systems for office-based additive manufacturing, rapid prototyping and direct digital manufacturing solutions. The company offers leasing service of some models of their manufactured commercial 3D printers and bundled 3D-printer packages in the United States. Besides the printer, the 3D Print Packs include start-up supplies, a support-removal system and cleaning agent. StrataSys also provides various separate services such as system operation, in-house support, education, project implementation and consulting (Figs. 12 and 13).

Fig. 12 StrataSys direct digital manufacturing solutions leasing, 2012

STRATASYS

1. Stratasys ships the printer pack (printer + printer support + cleaning products) to the customer office *

2. The customer pays for a monthly lease package (printer in his/her office + printer support + cleaning products)

3. Stratasys provides permanent support (when needed)

Enterprises/
Makers/
Designers/
Engineers

STRATASYS

FLOWS LEGEND Information ⊷ Labor ⟶ Finance ⟶ Materials ⟶   Product ownership   Partnership of providers

Fig. 13 StrataSys system map

Industree Foundation

Active since: 2000

Provider/s: Industree Foundation

Customers: Farmers, Artisans

S.PSS Type: Result Oriented

DE configuration: Decentralized Manufacturing

Products: Sustainable Producer Owned Enterprises

Services: Training enterprise leaders in business management, soft skills and hard skills, connecting to academia and designers, creating access to capital and markets, providing digital connectivity

Location: India, Ethiopia

The Industree foundation organizes rural communities in a distributed value chain yet integrated through an aggregated national level marketing and sourcing enterprise, with whom producer-owned enterprises have the choice to interact for some or all their transactions. The company holistically tackles the root causes of poverty by creating an ownership-based, organized creative manufacturing ecosystem for micro-entrepreneurs, most of whom are women. Distributed Design and Manufacturing that is equitable and sustainable can be viable only if there is an enabling ecosystem of support. Industree works to co-create an enabling platform using its 6C principle:

1. **Construct**: Business model innovation through producer ownership and inclusive entrepreneurship. Producer members earn through fair wages for production and shared profits from production and marketing. Aggregation for viability in material sourcing, professional management, productivity, access to market and capital.

2. **Capacity**: Training encompasses a grassroots business academy that trains producers and micro-enterprise leaders, paraprofessionals who work in the unit as professional support and service, and enterprise leadership. Training of professionals and enterprise leaders for broad handholding for replication and adaptation beyond Industree.

3. **Create**: Co-creation of design by professional designers who are part of the professional management group, alongside master artisans, designers would also be part of the professional management team. The efforts of the inhouse team will be bolstered by students and academic institutions from the region and beyond convened by Industree.

4. **Capital**: Creating access to capital through partnerships with Non-Banking Financial Company (NBFC) and working capital pools. A revolving working capital pool will be created along with funds offered through schemes of the Micro, Small and Medium Enterprises (MSME) sector and access to loans from banks based on purchase orders received.

5. **Channel**: Markets, connecting to markets both B2B and B2C, creating the awareness among buyers through meets and workshops, using brands to connect with

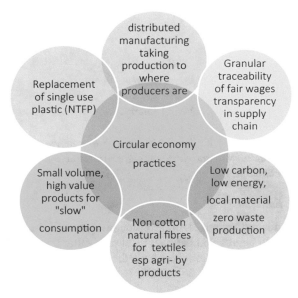

Fig. 14 An example of Circular Economy practices of Industree

brands. Participation in fairs and exhibitions to promote the products, nationally and regionally.

6. Connect: Digital connectivity primarily through mobile applications, through which sustainable enterprises in the creative manufacturing can be supported, serviced, incubated and accelerated (Figs. 14 and 15).

2.5 S.PSS and Distributed Software (DS)

While the internet began as a decentralized network of servers accessed by a network of users, there has now been a shift towards an increasingly centralized net through intermediaries like Google and Facebook whose servers handle a significant portion of all data on the internet. This has led to concerns over individual and organizational privacy, data protection and data ownership and agency. There is also a growing realization that these intermediaries have disproportionate control over information flows. Since online media now encompass critical sectors like finance, social networking and business, there are emerging alternatives that seek to develop networks of distributed and localized data storage and application embedded in communities rather than global corporates.

The case studies of S.PSS and Distributed Software (DS) presented below demonstrate a movement towards community-based and community-led online services. However, it is to be noted that although distributed software and in particular the

Fig. 15 Bangalore GreenKraft—one of the enterprises set up by Industree

case studies chosen show potential opportunities for developing S.PSS models, they have not yet actively incorporated it into their current form.

Secure Scuttlebutt

Active since: 2014

Provider/s: Secure Scuttlebutt

Customers: Community

S.PSS Type: Result-Oriented

DE configuration: Decentralized Software

Products: Offline Friendly Secure Gossip Protocol

Services: Data Ownership, End to End encryption, Agency over interaction

Payments:-

Location: Worldwide (origin New Zealand)

Secure Scuttlebutt is a localized but distributed social network that works with a peer-to-peer mesh network where user data is stored locally on user devices rather than a centralized server [7]. The data is exchanged between devices through data replication on a shared WiFi or local area network or even with a USB stick. It is also possible to connect to the network using public servers called "Pubs". The intent of Secure Scuttlebutt is to eliminate the need for connection to centralized servers while still having the network intact through a localized community of devices. This develops a resilient system that is upheld through the distributed network.

On a voluntary basis, it is possible for users to engage monetarily using the Secure Scuttlebutt Consortium. As an S.PSS model, in exchange for a voluntary donation, the developers are able to provide an opportunity for an alternate social media network that protects user data and allows the user to choose terms of engagement with the network. It connects people who do not have access to a regular internet service and can also be used in emergency situations.

Holochain

Active since: 2006

Provider/s: Holochain is a technology that can be used by multiple providers

Customers: Communities of users and developers

S.PSS Type: Product- or Use-oriented S.PSS (depending on the application)

DE conf.: Distributed Software

Products: HoloPort device for hosting (optional)

Services: Framework and protocol for app development

Payments: Hosts are paid in crypto; Holo takes a percentage transaction fee

Location: Worldwide

Holochain is a framework for building distributed peer to peer applications that is based on a shift from data centric computing of the Blockchain to an agent centric model. Holochain is a way of building and running applications on the user's own devices and without using an intermediary server. Users within a community that have spare computing capacity on their devices can host the applications of others. In exchange for this, the contributor gets paid in Holo Fuel, a crypto-currency that can be used to buy applications or hosting services within the community. Another characteristic of Holochain's agent-centric approach is that the users determine the terms of engagement within their own communities and this cannot be disrupted by others.

As a distributed computing system, parallels can be drawn with a two-way power grid, except here computing capacity is shared by users. It is possible to envision ways in which this peer-to-peer sharing system can allow users to build a sharing ecosystem of online and offline services in future (Fig. 16).

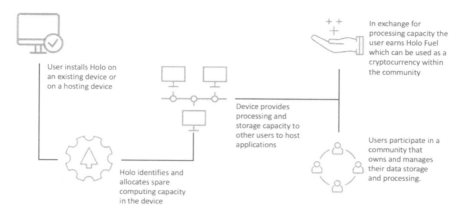

Fig. 16 Holochain system (Adapted from [5])

Linux

Active since: 1991

Provider/s: Community of developers

Customers: Community

S.PSS Type: Use-oriented S.PSS

DE conf.: Distributed Software

Products: Computer code

Services: -

Payments: Free

Location: Worldwide

Linux is an open-source operating system that is used in smartphones, personal computers, netbooks, supercomputers, servers, embedded devices, home appliances, cars and so on. The source code can be used, modified and distributed by anyone under the GNU General Public License, with the condition that whoever distributes software using a source code under the GNU license must make the original as well as the modified source code available under the same terms. Thus it can be said that Linux is software produced by a network of developers e.g. small-scale producers that are connected with each other locally and globally (Fig. 17).

Fig. 17 Examples of Linux application

2.6 S.PSS and Distributed Knowledge (DK)

Internet and web technology has revolutionized learning by providing a vast amount of learning resources across disciplines that are easily accessible and very affordable, often only at the cost of the internet service. One of the main ways in which this has influenced education in universities and schools is bringing ways of learning, thinking and doing to the forefront since it is now fairly common to be able to pick up skills through online resources. There is also a shift towards self-directed learning where students decide which subjects and skills are most appropriate to support their own goals and interests.

A number of online learning platforms like EdX and Coursera provide online courses from universities across the world and in a range of subjects that can be freely accessed by anyone with internet access. Additionally, on passing a course, it is also possible to pay a nominal fee for receiving a certificate from the respective university.

There are also attempts to draw a link between online learning communities and physical communities in distributed knowledge production and application. The case study of S.PSS and Distributed Knowledge (DK) in this section illustrates one such attempt.

Project DEFY (Design Education for Yourself)

Provider/s: Project DEFY

Customers: Community

S.PSS Type: Use-oriented S.PSS
DE configuration: Decentralized Knowledge

Products: Nooks

Services: Induction Program for new learners

Payments: Income generation through innovation and projects

Location: India, Uganda, Rwanda, Zimbabwe

Location: India

Project DEFY sets up self-designed learning centres or 'Nooks' across marginalized communities in India and Africa (Uganda, Rwanda, Zimbabwe). Nooks are free-for-all 'schools without teachers' that provide everyone in the local community with access to technology, tools, resources and information to design their own education. As such, Nooks are a primary example of distributed education design where learning is decentralized, contextualized, localized and individualized.

This process is supported through a 45-day long induction programme for new learners in which they get exposed to new areas of skills and learning through hands-on practice as well as through fostering and providing a safe, inclusive space for meaningful conversations to take place among the Nook community. At the end of the induction programme, the learners are enabled (individually or in groups) to identify and write down their own, individual goals and help to break them down into concrete, hands-on projects they pursue in order to achieve their goals.

As opposed to schools and colleges where decision-making follows an authoritarian top-down approach, Nooks are managed by the community members themselves. This includes administrative decisions such as the opening times of the Nook, the responsibility for a monthly resources budget, as well as—of course—being in charge of the learning process itself.

Whereas in schools and colleges the learners are separated from the means and resources of learning—having their relations to those means mediated, appropriated, circumscribed and severed by teachers, textbooks, curriculum, etc.—Nook learners are enabled to take control of and directly own the means and resources

Fig. 18 Nook enabled by Project DEFY (Source: Project DEFY)

of learning. Importantly, Nooks do this on a scale and cost that fits within the economy of even low-income communities. In the long run, the Nooks aim to become self-sufficient and completely community-run by capitalizing on the creations, innovations, products and skills that emerge out of them (Fig. 18).

2.7 S.PSS and Distributed Design (DD)

The complex nature of contemporary design problems has led towards an increasingly collaborative and heterogeneous approach to knowledge production and design. There is a recognized need for bringing together experts and stakeholders across disciplines to fully understand the nature of dependencies of a system and to then design appropriate systemic solutions [4, 12].

A number of design platforms and collectives have emerged over the last decade. Some are along the lines of an Uber model, where designers, manufacturers and suppliers are connected with customers to enable distributed design service. In disciplines like architecture and interior design, for example, multiple companies have emerged which provide end-to-end services of design, fabrication, installation and finish with additional options such as home products and maintenance services using distributed networks of local businesses. Other distributed design models such as Local Motors are more topical and specific.

Several case studies of S.PSS and Distributed Design (DD) are presented below.

Quirky

Active since: 2009

Provider/s: Quirky (platform, online tool and connection between members and manufacturers), Partners, e.g. General Electric, PepsiCo, Mattel (manufacturing)

Customers: Designers, inventors, individuals with specific skills

S.PSS Type: Result-oriented S.PSS

DE configuration: Distributed Design

Products: -

Services: Provides a network of skilled users and access to product creation enterprises

Payments: Free (use of the platform)

Location: New York City.

Quirky is an invention platform that connects inventors with users who have other skills for developing the idea and with companies specialized in a specific product category for manufacturing. The offer is therefore access to complete product creation. Quirky's business model pays designers part of the profits that their products yield. The users do not need to pay for using the platform. The users can submit their ideas and connect with others to make a team for collaboration. Once the developed idea is accepted by Quirky through a voting system by the Quirky community at Eval (Quirky's live weekly product evaluation), it is pitched to the manufacturers. If it is manufactured, Quirky shares the profit with the team members according to their influence evaluated by a point system on the Quirky platform (Fig. 19).

Local Motors

Active *since: 2007*

Provider/s: Enthusiasts, hobbyist innovators, designers, engineers, fabricators and other professionals

Customers: Designers, inventors, individuals with specific skills

S.PSS Type: Hybrid of use-oriented and product-oriented S.PSS

DE configuration: Distributed D and Decentralized Manufactoring

Products: Motor Vehicles (rally cars, motorcycles, electric bicycles, tricycles, children's ride-on toy cars, radio-controlled model cars and skateboards)

Fig. 19 Quirky invention platform, 2009

Fig. 20 Local Motors vehicle manufacturing company, 2007

Services: Management of the global network of microfactories and the co-creation community

Payments: Free (users get even paid in case of revenue for their contribution)

Location: USA

Local Motors (LM) is a motor vehicle manufacturing company focused on low-volume manufacturing of open-source motor vehicle designs using multiple micro-factories. Their products include the Rally Fighter automobile and Racer motorcycle, various electric bicycles, tricycles, children's ride-on toy cars, radio-controlled model cars and skateboards. They 3D print some components. Rally Fighters have used co-creation techniques, whereby products are designed cooperatively with end-users, as part of its designing phase. Their website is a community focusing on engine vehicle innovation. The content is co-created by the users of the community who discuss designing, engineering and building innovative engine vehicles (Fig. 20).

3 S.PSS Applied to DE: A Scenario

Envisioning the coupling of the two offer models, S.PSS and DE, some further considerations highlighting some of their evident characteristics could be given.

First of all, as far as we have diverse types of DE (Distributed energy Generation (DG), Distributed Manufacturing (DM), Distributed production of Food (DF), Distributed Water management (DW), Distributed production of Software (DS), Distributed production of Information/knowledge (DI), Distributed Design (DD) and the 3 main types of S.PSS (Product-oriented S.PSS, Result-oriented S.PSS; Use-oriented S.PSS), it is clear that a set of diverse combinations of these could emerge in principle.

Secondly, another possible main variable is the type of customer or user, i.e. whether a B2B offer, B2C offer, p2p non-market, and so on.

Furthermore, as far as the **hardware** of each DE and who is producing it is a key characteristic (because in an S.PSS offer model she/he is the one that has the economic interest to redesign it with a low environmental impact), it is of key interest to identify the hardware for each type of DE.

Finally, other characteristics worth highlighting are related to the DE structure types, i.e. they could be **Distributed** or **Decentralized** and each of those could be **stand-alone** or **network-structured**.

In relation to those variables and their possible sustainable combinations, a *Sustainable Design-Orienting Scenario* (SDOS) for *Sustainable Product-Service System (S.PSS)* applied to *Distributed Economies (DE)* in low and middle-income (all) contexts has been designed[1] to provide a new vision of sustainable production & consumption systems.

---

[1]The scenario design process emerged from a case study analysis of S.PSS applied to DE (best practices), as well as an idea generation workshop focused on S.PSS applied to DE using the SDO toolkit (www.lens-international.org). The scenario presented here is an update of a scenario

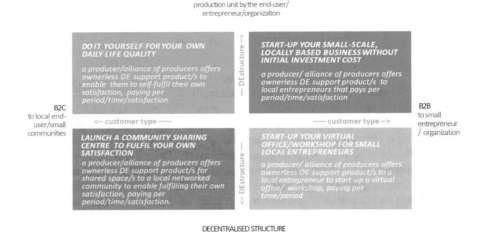

Fig. 21 The Sustainability Design-Orienting Scenario for S.PSS applied to DE

The Scenario is composed of a polarity diagram with 4 visions, for each of the 4 quadrants in the diagram matrix. Each vision represents a Sustainable win-win configuration, combining socio-cultural, organizational and technological factors, fostering solutions with a low environmental impact, a high socio-ethical quality and a high economic value.

The scenario matrix is polarized on the vertical axis by the type of DE structure, distributed or decentralized, and on the horizontal axis by the type of customer/user, B2C (final user or small communities) or B2B (small entrepreneur or small business). The crossing of those polarities produced the following 4 visions, relative to the four quadrants (see Fig. 21)[2]:

A. **[distributed-B2C] DO IT YOURSELF FOR YOUR OWN DAILY WELL-BEING**: a producer/alliance of producers offers ownerless DE support product/s to enable the end-user to self-fulfil their own satisfaction, paying per unit of period/time/satisfaction.

B. **[distributed-B2B] START-UP YOUR SMALL-SCALE, LOCALLY BASED BUSINESS WITHOUT INITIAL INVESTMENT COST**: a producer/alliance of producers offers ownerless DE support product/s to local entrepreneurs that pay for unit of period/time/satisfaction.

C. **[decentralized-B2C] LAUNCH A COMMUNITY SHARING CENTRE TO FULFIL DAILY LIFE SATISFACTION**: a producer/alliance of producers

---

developed by Cenk Basbolat for his degree thesis at the School of Design of Politecnico di Milano, tutored by Carlo Vezzoli.

[2]A set of videos presenting the visions of the scenario as well as their possible options are available at http://lens-europe.eu/tools/view/2

offers ownerless DE support product/s for shared space/s to a local networked community to enable fulfilling their own satisfaction, paying per period/time/satisfaction.

D. **[decentralized-B2B] START-UP AS SMALL, LOCAL ENTREPRENEURS WITH VIRTUAL OFFICE/WORKSHOP**: a producer/alliance of producers offers ownerless space for an office and/or workshop equipped with DE support product/s to a local entrepreneur to start-up its business, paying per unit of time/period.

To illustrate the scenario, we now describe one example (case study) per each of the visions.

A. Do-it-yourself for your own daily life quality: example

Qurrent: *The company teaches customers how to produce and manage renewable energy, allowing them to organize the exchange of energy in small local networks. Qurrent offers Solar Home Systems (SHS) composed of photovoltaic panels (and related components) and three core products: Qbox, Qmunity website, Qserver. Specifically, the Qbox measures all production and consumption of electricity and makes it possible to share capacities with the neighbourhood.*

B. Start-up your small-scale, locally-based business without initial investment cost: example

SELCO: *With the support of government funds and bank loans, SELCO facilitates and enables financing options for rural entrepreneurs to set up solar powered enterprises like photocopying and printing kiosks, tailoring units and mechanized cattle milking units in underserved areas. This generates sustainable livelihood options and offers access to services for the community.*

C. Launch a community sharing centre to fulfil daily life satisfaction: example

Helsinki Metropolitan Area Reuse Centre: *The Reuse Centre sells donated second-hand goods and building and hobby materials in their retail outlets, which are located in many locations in Helsinki, Finland. The organization makes it easier for customers to reuse and recycle by offering transportation services and leasing pull-trailers for a fee, and customers can borrow a cargo-bike for free if they purchase something from one of the shops. The Reuse Centre also provides educational workshops on recycling and the environment to children.*

D. Start-up a virtual office/workshop for small, local entrepreneurs: example

> *Maker Station*: *Maker Station is a large makerspace, workshop and co-working space that provides access to industrial tools and equipment, studio space and storage space for artists, artisans, designers and small producers on a membership basis in Cape Town, South Africa. It also links artists, artisans and designers with projects and companies needing their skills and labour, and it provides technology workshops for hobbyists and marginalized children. The equipment and tools it provides include milling machines and lathes, laser cutters, CNC vinyl cutters, 3D printers, electronics stations, sheet metal equipment, welding equipment, woodworking equipment, hand tools and sewing machines.*

Exploring opportunities within the S.PSS applied to DE Scenario

Within the SPSS applied to DE Scenario, the following strategies (and guidelines) have been identified[3] as potential diversification of proposals within each of the visions:

- Complement DE hardware offer with Life Cycle services
- Offer ownerless DE systems as enabling platform
- Offer ownerless DE systems with full services
- Optimize stakeholders' configuration
- Delink payment from hardware purchases and resource consumption
- Optimize DE structure.

Complement DE hardware offer with Life Cycle services

- The **provider/s** complements the offer of the **DE** system with:

  – **financial services** to support **initial investment** and eventual maintenance and repairing costs, e.g. micro-credit, crowdfunding, donation to **maintain**, **repair** one or more **DE hardware/components**
  – support **services** for the **design** and/or **installation** of its components (e.g. in DG, the micro-generator, the storage, the inverter, the wiring, etc.)
  – support **services during use**, i.e. maintenance, repairing and upgrading of its components
  – support **services** for the **end-of life treatment** of its components
  – support **services** to enable the customer to either **design** and **produce with their DE hardware**, **share** their **DE** hardware, **sell/provide their** production, **provide services** through their DE hardware.

---

[3]Those presented here are an update of a set of criteria and guidelines developed by Cenk Basbolat for his degree thesis at the School of Design of Politecnico di Milano, tutored by Carlo Vezzoli.

Offer ownerless DE systems as enabling platform

- The **provider/s** complements an ownerless offer of the **DE** system with **training/information** services **to enable** the customer:
  - to **design** the DE **hardware/components**
  - to **maintain**, **repair** one or more **DE hardware/components**
  - to **install** one or more **DE hardware/components**
  - to **upgrade** one or more **DE hardware/components**
  - to **optimize use** of one or more **DE hardware/components**
  - to either **design**, **produce with their DE** hardware, **share** their **DE** hardware, sell/provide their products, provide services through their **DE** hardware.

Offer ownerless DE systems with full services

- The **provider**/s complements an ownerless offer of the **DE** system with **full** support **services**:
  - to **design** the **DE hardware/components**
  - to **maintain**, **repair** one or more **DE hardware/components**
  - to **install** one or more **DE hardware/components**
  - to **upgrade** one or more **DE hardware/components**
  - to **optimize use** of one or more **DE hardware/components**
  - to either **design**, **produce with their DE** hardware, **share** their **DE** hardware, **sell/provide** their production, **provide services** through their **DE** hardware.

Optimize stakeholders' configuration

- Offer the S.PSS to the final user, or a collective, to **improve the quality of life** or the environment
- Offer the S.PSS to an entrepreneur to **enable a business start-up** or empower business
- Optimize a stakeholder partnership with **vertical integration** by combining all complementary components of one single DE type (e.g. in DG, the micro-generator, the storage, the inverter, the wiring, etc.)
- Optimize a stakeholder partnership with **horizontal integration** (by combining different DE offers as a full package offer)
- Make the DE hardware manufacturer S.PSS offers either alone or in a joint venture with another stakeholder
- Make the DE service provider S.PSS offers either alone or in a joint venture with another stakeholder.

Delink payment from hardware purchases and resource consumption

- Offer **pay x period**, i.e. the cost is daily/weekly/monthly/yearly fixed
- Offer **pay x time,** i.e. the cost is fixed per minutes/seconds of access
- Offer **pay x use**/satisfaction unit, i.e. the cost is fixed per product performance (e.g. km for a vehicle, washing cycles for a washing machine)
- Offer payment based on **hybrid** pay x period, pay x time, pay x use modalities

- Offer other economic transactions not based on financial currencies, such as time exchange or direct exchange of goods
- Apply for additional financial support from public administrations/entities.

Optimize DE structure

- Offer **stand-alone DE** Product-Service Systems for homes or business sites (especially isolated sites)
- Offer local **mini-network** connecting **DE** systems, to enable local production surplus sharing or for enabling shared use of the **DE** hardware and sharing operations for DE service provision
- Offer **decentralized stations**, e.g. 3D printing service spot, charging spot, etc., for local communities or decentralized service providers, e.g. a local technician's shop
- Offer **decentralized** systems to locally supply DE production throughout a **mini-network** for homes and/or business sites or a mini-network of service providers
- Offer the DE system with a **connection** to a **worldwide network/main-grid**, enabling homes, small business and communities the selling/purchasing of production or for enabling shared use of the DE hardware and/or the shared provision of local services.

4   Barriers and Trade-Offs to Integrating S.PSS and DE

Distributed Economies evolve with context and situation to provide different opportunity spaces that can attract new stakeholder configurations. This requires a constant adjustment of the S.PSS model, which can be informed by these changing opportunities. This means that the organization needs to be flexible in allowing possibility for customization according to local needs and context of different network nodes in terms of providing a relevant S.PSS model.

As evidenced by many of the case studies in this chapter, setting up these systems needs long-term engagement with local communities and networks. It must include capacity building, community mobilization and awareness creation at multiple nodes. Bringing together these capabilities and developing networks for learning and sharing the know-how. However, we must have sustained and long-term investment in developing the knowledge and capacity in multiple regional and local centres, and this can be a challenging proposition.

Scaling up of distributed networks again can be challenging, as it requires strategic components to be centralized to optimize resources. Finding the balance between the distributed and centralized components of a product-service system in a distributed economy must include an intelligent business plan backed by policies that support the sustainability of the system.

Distributed Economies may not necessarily provide the most environmentally efficient solution. It is important to assess and balance all three components of sustainability—economic, environmental and socio-ethical—to arrive at the best possible model. This would be possible if we had increased access to assessment tools and frameworks as well as modelling technology, all of which today are not very easily available to planners and entrepreneurs at the grassroots level.

References

1. Beck MW, Claassen AH, Hundt PJ (2012) Environmental and livelihood impacts of dams: common lessons across development gradients that challenge sustainability. Int J River Basin Manag 10(1):73–92
2. Emili S, Ceschin F, Harrison D (2016) Product-service systems applied to distributed renewable energy: a classification system and 15 archetypal models. Energy Sustain Dev 32:71–98
3. Emili S (2017) Designing product-service systems applied to distributed renewable energy in low-income and developing contexts: a strategic design toolkit. PhD Thesis, Brunel University London
4. Giaccardi E, Fischer G (2008) Creativity and evolution: a metadesign perspective. Digital Creativity 19(1):19–32
5. Marin F (2018, March 8). Fixing the internet: how holochain wants to change the way we interact. *Ouishare Magazine*. https//www.ouishare.net/article/fixing-the-internet-how-holochain-wants-to-change-the-way-we-interact. Accessed 22 June 2020
6. Reinders A, Diehl JC, Brezet H (2012) The power of design: product-innovation in sustainable energy technologies. Wiley, London, UK
7. Tarr D, Lavoie E, Meyer A, Tschudin C (2019) Secure Scuttlebutt: an identity-centric protocol for subjective and decentralized applications. In: Proceedings of the 6th ACM conference on information-centric networking, pp 1–11
8. Vasantha GVA, Roy R, Lelah A, Brissaud D (2012) A review of product–service systems design methodologies. J Eng Des 23(9):635–659
9. Vezzoli C, Ceschin F, Diehl JC (2015) The goal of sustainable energy for all. J Clean Prod 97:134–136
10. Vezzoli C, Ceschin F, Osanjo L, M'Rithaa MK, Moalosi R, Nakazibwe V, Diehl JC (2018) Designing sustainable energy for all: sustainable product-service system design applied to distributed renewable energy. Springer, London
11. Vezzoli C, Kohtala C, Srinivasan A, Diehl JC, Fusakul M, Xin L, Sateesh D (2014) Product-service system design for sustainability. Greenleaf, Sheffield
12. Zha XF, Du H (2006) Knowledge-intensive collaborative design modeling and support: Part I: review, distributed models and framework. Comput Ind 57(1):39–55

Designing S.PSS and DE: New Horizons for Design

Carlo Vezzoli, Aine Petrulaityte, Sharmistha Banerjee, Pankaj Upadhyay, and Ravi Mokashi Punekar

1 General Considerations for Conceptual Integration into the Design Process

Assuming S.PSS applied to DE is an opportunity for a locally based sustainability for all, as introduced in this volume, we envision a new role for designers:

Designing Sustainable Product-Service Systems applied to Distributed Economies, or shortly **System Design for Sustainability for All (SD4SA)**.

To introduce this topic, the following preliminary definition could be given to articulate the new potential of such design:

> System Design for Sustainability for All (SD4SA):
>
> design of S.PSS applied to DE, i.e. the design of Systems of Products and Services that are together able to fulfil a particular customer demand (deliver a "unit of satisfaction"), within the Distributed Economies paradigm; based on the design of innovative interactions among locally-based stakeholders, where the ownership of the product/s and/or its life cycle responsibilities/costs remain with the provider/s, so that the provider/s continuously seek environmentally and/or socio-ethically beneficial new solutions accessible to all with economic benefits.

Within this framework a new knowledge-base and know-how emerge: competences in designing and implementing Sustainable Product-Service Systems applied to Distributed Economies, i.e. Distributed energy Generation, Distributed production

C. Vezzoli
Design Department, Politecnico di Milano, Milan, Italy

A. Petrulaityte (✉)
Department of Design, Brunel University London, London, UK
e-mail: aine.petrulaityte@brunel.ac.uk

S. Banerjee · P. Upadhyay · R. M. Punekar
Indian Institute of Technology Guwahati, Guwahati, India

© The Author(s) 2021
C. Vezzoli et al. (eds.), *Designing Sustainability for All*, Lecture Notes
in Mechanical Engineering, https://doi.org/10.1007/978-3-030-66300-1_4

of Knowledge, Distributed Software development, Distributed Manufacturing and Distributed Design.

Based on the foundations of S.PSS design [10, 6], the following approaches and skills can be identified and refined for System Design for Sustainability for All (SD4SA):

(a) "Satisfaction-system" approach: This calls for skills to design the system of products and services that, within a DE paradigm, can satisfy a particular demand ("satisfaction unit");
(b) "Stakeholder configuration" approach: This calls for skills to design the stakeholders' interactions in a particular DE satisfaction-system;
(c) "System sustainability4all" approach: This calls for skills to design-for-all a DE system where the providers continuously seek environmental and/or socio-ethical beneficial new solutions, with economic benefits.

This new role in SDS4A calls for the design of "appropriate stakeholder configurations" and favouring the design of "appropriate technologies", to address S.PSS applied to DE. In this framework, two key approaches have been merged, redefined and updated: Product-Service System Design for Sustainability and Distributed Economies (DE) design. Other disciplines that are not explicitly mentioned here could and should also be included, to contribute to a comprehensive and complete research base, e.g. Social Entrepreneurship for Sustainable Development and System Innovation for Sustainability.

2 A Reference Model of S.PSS and DE Design

2.1 Method and Tools for System Design for Sustainability for All

Criteria, method and tools

Before introducing and describing methods and tools, let us summarize the main issues discussed so far. It has been argued that a potential role exists for design for sustainability, in promoting and facilitating innovation resulting in environmentally beneficial, economically viable and socially equitable/cohesive enterprises/initiatives offering a mix of products and services, especially when applying the Sustainable Product-Service Systems (S.PSS) model to Distributed Economies (DE).

The **first key point** is the approach to design a stakeholders' configuration, which is committed to creating and promoting innovative types of interactions and partnerships between appropriate socio-economic stakeholders of a system responding to a particular social demand (unit of satisfaction). Consequently, new skills are required from the designer, directly or as a facilitator of a design process:

- A designer must be able to design both products and services, related to a given demand (needs and/or desires), i.e. a satisfaction unit;
- A designer must be able to identify, promote and facilitate innovative configurations (i.e. interactions/partnerships) between and among different stakeholders (entrepreneurs, users, NGOs, institutions, etc.), i.e. a satisfaction system related to a given demand (needs and/or desires) as a satisfaction unit

The **second key point** emphasizes S.PSSs applied to DE innovations that are environmentally, socio-ethically and economically sustainable, i.e. they have a low environmental impact and promote socially equitable and cohesive results, with economic benefits. This underlines that the design process should be oriented towards sustainable solutions, i.e. a designer must be capable to design S.PSSs applied to DE systems (and related stakeholder interactions) that couple economic benefits with environmental and socio-ethical, beneficial new solutions. Consequently, these new skills are required from the designer:

- The ability to orientate the system design process towards *eco-efficient* solutions, encompassing both environmental and economic sustainability;
- The ability to orientate the system design process towards *socio-efficient* solutions encompassing both socio-ethical and economic sustainability.

In this chapter, we describe a series of tools that can be applied during different phases of a design process. Besides their specific functions, more generally, they are meant to assist the designer in accomplishing three objectives:

1. Assessing existing system sustainability and defining sustainability system design priorities;
2. Generating a sustainability-focused system idea and concept (innovative S.PSS applied to DE);
3. Checking/visualizing the sustainability improvement/worsening of developed system concept/s (comparing the existing baseline with the new, innovative system).

Various research projects have been funded by the European Union and the United Nations Environment Programme (UNEP)[1] over the past decades to develop and test methods and tools for system design, the main ones being SusHouse,[2] ProSecCo,[3] HiCS,[4] MEPSS,[5] SusProNet,[6] LeNS[7] and LeNSes.[8]

---

[1] Design for Sustainability (D4S): A Step-By-Step Approach (UNEP funded, 2005–2009) (see [6]).

[2] SusHouse: Strategies towards the Sustainable Household (EU funded, 1998–2000) (see [9]).

[3] ProSecCo: Product-Service Co-design (EU funded, 2002–2004).

[4] HiCS: Highly Customerised Solutions (EU funded, 2001–2004) (see Manzini et al. [3]).

[5] MEPSS: MEthodology for Product Service System development (EU funded, 2002–2005) (see [8]).

[6] SusProNet: Sustainable Product-Service co-design Network (EU funded, 2002–2005) (see [7]).

[7] LeNS: Learning Network on Sustainability (2008–2010).

[8] LeNSes: Focused on System Design for Sustainable Energy for all (EU-funded, Oct 2013–Oct 2016).

In this chapter, the Method for System Design for Sustainability (MSDS) is described, together with its tools for System Design for Sustainability for All, i.e. design of S.PSS applied to DE. It is important to note that experimentation, both in applied research projects and in teaching, has been fundamental and will continue to be so in future, in order to allow methods and tools to be assessed, adapted and improved.

2.2 MSDS: A Modular Method for System Design for Sustainability

The MSDS method aims to support and orient the entire process of system innovation development towards sustainability. It is conceived for designers and companies, but it is also appropriate for public institutions, NGOs and other types of organizations. It can be used by an individual designer or by a more extensive design team. In all cases, special attention has been paid to facilitate the co-designing processes both within the organization itself (between people from different disciplinary backgrounds) and outside, bringing different socio-economic actors and end-users into play. The method is organized in stages, processes and sub-processes. It is characterized by a flexible modular structure so that it can easily be adapted to the specific needs of designers/companies/organizations and to diverse design contexts and conditions. Its modular architecture is of particular interest in terms of the following considerations:

- Stages/processes: all stages and related processes can be undertaken, or only some depending on the particular requirements of the project;
- Tools: the method is accompanied by a series of tools that can be selected and deployed during the design process according to the project need;
- Dimensions of sustainability: the method takes into consideration the three dimensions of sustainability (environmental, socio-ethical and economic). It is possible to choose which dimension(s) to operate on;
- Integration of other tools and activities: the method is structured in such a way as to allow the inclusion of design tools that have not been specifically developed for it. It is also possible to modify existing activities or add new ones according to the particular aspects of the design project.

The basic structure of *MSDS* consists of four main stages:

1. Strategic analysis
2. Exploring opportunities
3. System concept design
4. System detailed design.

An additional stage can be added, across the others, for drawing up documents to report on the sustainability characteristics of the designed solution:

- Communication.

Table 1 shows the aim and processes for each stage.

Table 1 The stages of MSDS with their aims and processes. Sustainability-oriented processes are in bold

MSDS method		
Stage	Aim	Processes
1. Strategic analysis	To obtain the necessary information to facilitate the generation of sustainable system innovation ideas	• Analyse the project proposers and outline of the intervention context • Analyse the context of reference • **Analyse the carrying structure of the system** • **Analyse cases of sustainable best practice** • **Analyse the sustainability of the existing system and determine priorities for the design intervention in view of sustainability**
2. Exploring opportunities	To make a 'catalogue' of promising strategic possibilities available: a sustainability design-orienting scenario and/or a set of promising sustainable system ideas	• Benchmark against sustainable solutions for similar problems • **Generate sustainability-oriented system ideas** • **Outline a design-oriented sustainability scenario**
3. System concept design	To develop one or more system concepts oriented towards sustainability	• Select clusters and single ideas • Develop system concepts • **Conduct environmental, socio-ethical and economic assessment** • Visually represent the most promising concept
4. System detailed design (and engineering)	To develop the most promising system concept into the detailed version necessary for its implementation	• Detailed system design • **Review environmental, socio-ethical and economic issues and visualization**
5. Communication	To draw up reports to communicate the general and above all sustainable characteristics of the system designed	Draw up the documentation for communications of sustainability

3   Tools Developed by LeNS

This section describes several tools that may be used to support the various stages of the *Method for System Design for Sustainability* (MSDS) with an integration of Distributed Economies (DE).[9] In general, the tools are created to support designers to achieve four objectives:

- To assess existing systems and define sustainability design priorities;
- To explore opportunities by generating sustainability-oriented system ideas with a specific focus on S.PSS applied to DE;
- To visualize the proposed S.PSS and DE concept design;
- To detail and communicate the proposed S.PSS and DE concept design by highlighting environmental, social and economic benefits.

Seven S.PSS and DE design support tools, newly developed within the LeNSin project, are presented below:

- Sustainability Design-Orienting Scenarios (SDOS) on S.PSS and DE

- Innovation Diagram for S.PSS and DE

- Concept Description Form for S.PSS and DE

- System Map for S.PSS and DE

- S.PSS and DE Idea Borads (embedded into the SDO toolkit)

- Strategic Analysis Toolkit (SAT) for DE for Socio-Economic Ecosystems (SEE)

- Distributed Manufacturing (DM) applied to PSS design toolkit.

MSDS and other tools for system design for sustainability have been developed to support system design for sustainability for all, and a wide selection of these tools can be found and downloaded from the LeNS platform (www.lens-international.org). This particular section of the book aims to help potential users to apply the newly developed S.PSS and DE tools in practice. For this reason, each tool is described using the following structure:

1. The aim and the components of the tool;
2. The tool's integration into the MSDS design process;
3. How to use the tool;
4. Availability and resources required.

---

[9]This is the one of the outcomes of the LeNSin project, creating, integrating and updating tools produced by the project partners together with other existing tools and approaches linked to system design for sustainability. The tools described here are a selection of those that have been used and tested during a set of pilot courses as part of the LeNSin project and in several studies with companies and industry experts.

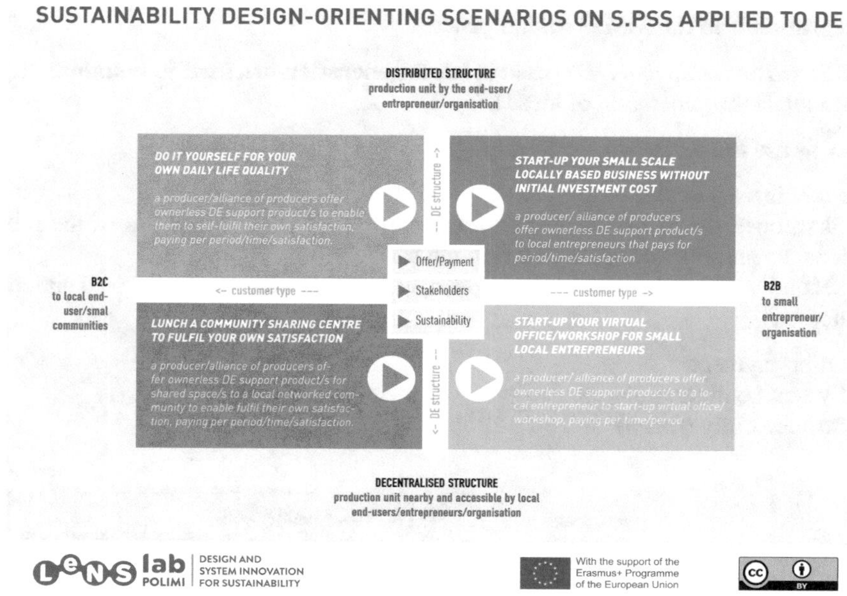

Fig. 1 SDOS on S.PSS and DE tool

3.1 Sustainability Design-Orienting Scenarios (SDOS) on S.PSS and DE

Aims

The objective of *Sustainability Design-Orienting Scenarios (SDOS) on S.PSS and DE* (Fig. 1) is to orient the design process towards sustainable system solutions by using immersive and inspiring scenario videos to stimulate the generation of *S.PSS-based DE* ideas for all.

Components

The *Sustainability Design-Orienting Scenarios on S.PSS and DE* consist of:

- Four main visions' videos
- Three sub-videos presenting options for all the visions in terms of:

  - Offer/payment
  - System configuration
  - Sustainability.

Integration into the MSDS design process

The *SDOS on S.PSS and DE* is used in **Ideas generation oriented to sustainability** to stimulate the generation of ideas (Fig. 2).

How to use the *SDOS on S.PSS and DE*

The tool is used in two simple steps:

First, open the *SDOS on S.PSS and DE* tool. Play the four videos of the four visions, to get initial design inputs through sample stories (Fig. 3).

Secondly, play the three sub-videos, to open up sample stories linked to all options related to:

• Offer/payment
• System configuration
• Sustainability (Fig. 4)

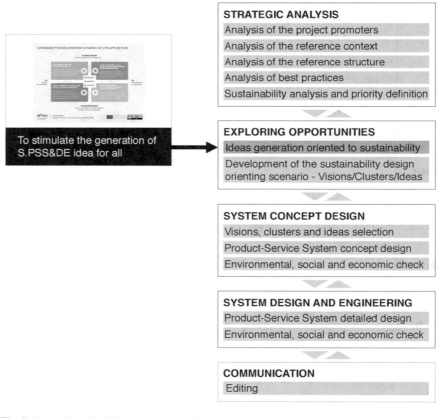

Fig. 2 Integrating SDOS Scenario on S.PSS and DE into the MSDS process

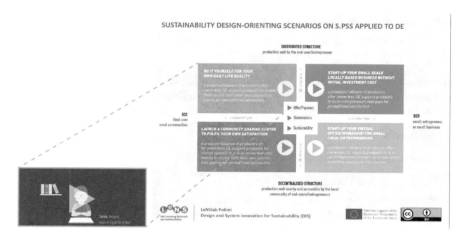

Fig. 3 The menu page of the SDOS on S.PSS and DE tool with the links to the four vision videos

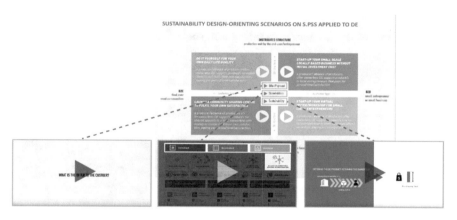

Fig. 4 SDOS on S.PSS and DE tool links to the three sub-videos visualizing offer/payment, system configuration and sustainability

Availability and requested resources

The *SDOS on S.PSS and DE* tool is an open-access tool that can be downloaded for free from www.lens-international.org, 'Tools' section. A computer, a PDF reader and Internet connection are required to access the tool.

The tool may be used by a single designer, though the support of a multi-disciplinary team is preferable.

This tool requires 15 min to explore and get inspired by the proposed visions.

3.2   Innovation Diagram for S.PSS and DE

Aims

The objective of the *Innovation Diagram for S.PSS and DE* (Fig. 5) is to help designers to position and characterize existing offers and competitors and select promising ideas for new concept profiling.

Components

The diagram consists of:

- Polarity diagram concept profile
- Digital sticky notes
- Database of labels.

Integration into the MSDS design process

The Innovation Diagram for S.PSS and DE is used at various stages of the design process (Fig. 6).

- In **Analysis of the project promoters and the reference context,** it can be used to:

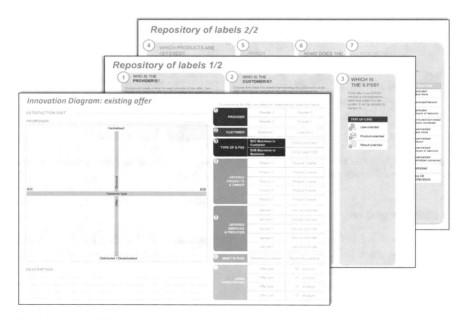

Fig. 5 Innovation Diagram for S.PSS and DE

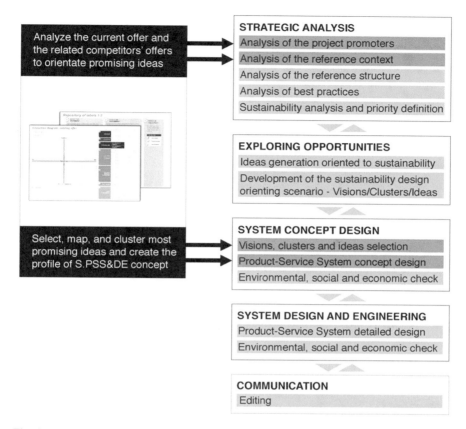

Fig. 6 Integrating Innovation Diagram for S.PSS and DE integrated into MSDS design process

- Analyse the current offer and the related competitors' offers to orientate promising ideas.

- In **Visions, clusters and ideas selection and System concept development** it can be used to:

   - Select, map and cluster the most promising ideas and create the profile and S.PSS and DE concept.

How to use the *Innovation Diagram for S.PSS and DE*

First, open the tool and move to the "..._existing offer" slide. Work in the "existing offer" slide to position an existing offer (Fig. 7). Select the company/organization icon (1) and choose one of the DE types to substitute the general one. Paste the label in the diagram and write the company/organization name in the free space on the label.

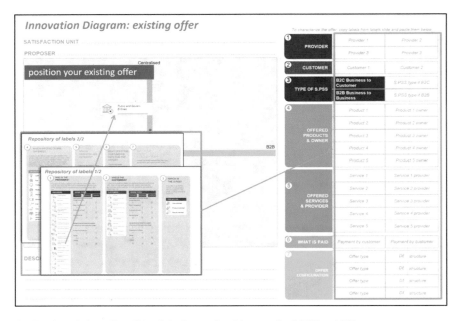

Fig. 7 The existing offer slide of the Innovation Diagram for S.PSS and DE

The second step is to characterize the existing offer by specifying all of the following (Fig. 8):

- **Provider (1)**. Select the company/organization structure label, and choose one of the Offer type characterization icons (Distributed, Decentralized or Centralized and its sector, i.e. energy Generation, Food production, Water management, Manufacturing, Software development or Knowledge) to substitute the general one. Place in the provided section and write the company/organization name in the free space on the label.
- **Customer (2)**. Select customer/s (B2B–B2C) structure label icon/s and choose one of the characterization icons to substitute the general one. Place in the customer section and write customer/s name in the free space on the label.
- **Type of PSS (3)**. Select the S.PSS type of the offer (if any): PRODUCT-ORIENTED, USE-ORIENTED, RESULT-ORIENTED and place it in the S.PSS type section. Remember, that in most cases existing offers are not S.PSS.
- **Offered Products & owners (4)**. Select the product icon representing what the company offers and paste in the products section. Select who retains the product OWNERSHIP (provider or customer) and place the label in the provider/customer label.
- **Offered Services & providers (5)**. Select the service icon representing what the company offers and paste in the service section. Select who PROVIDES the service and place the label in the provider label.

Fig. 8 Elements to characterize in the existing offer

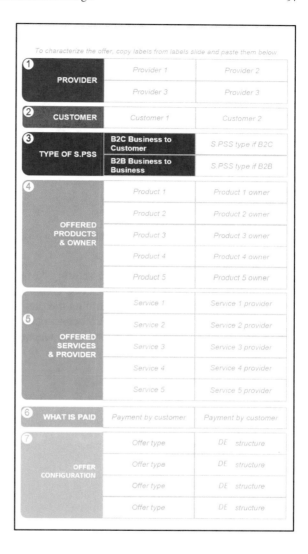

- **What is paid** (6). Select the icon describing what is paid by the customer/s and place the label in the payment section.
- **Offer configuration** (7). Select the DE type icon of the offer and paste it in the DE type space. Select its structure icon and place it in the nearby space.

The same process to characterize the existing offer could be done in relation to competitors, by moving to the "..._Competitors" slide. Finally, the Innovation Diagram could be used to insert and position promising ideas designed with the idea boards (SDO toolkit), within the "..._Concept" slide (Fig. 9).

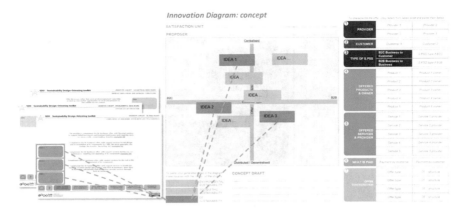

Fig. 9   The concept slide of the Innovation Diagram for S.PSS and DE with the SDO idea boards

Now it is time to generate new ideas spotting the areas that are left empty (Fig. 10). Identify and cluster those ideas that can be combined to draft the system concept. Write a text (max 200 characters) outlining the preliminary system concept.

Finally, profile an S.PSS and DE draft concept by copying and pasting characterizing icons of the emerging S.PSS and DE concept (Fig. 11).

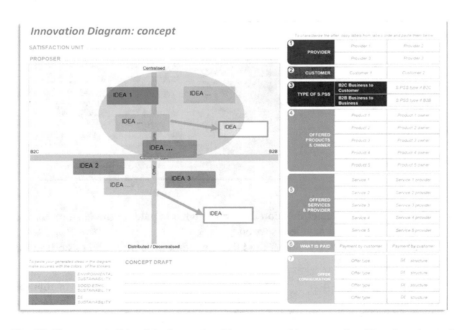

Fig. 10   The concept slide of the Innovation Diagram: new idea generation, idea clustering and system concept drafting

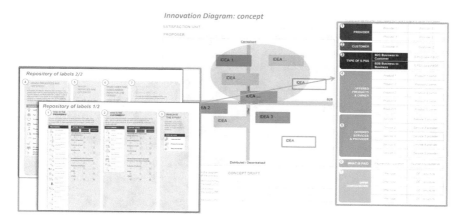

Fig. 11 Innovation Diagram: profiling an S.PSS and DE draft concept

Availability and requested resources

The Innovation Diagram for S.PSS and DE is an open access tool that can be downloaded for free from www.lens-international.org, 'Tools' section. A computer and a PowerPoint reader are needed to access the tool. This tool requires at least:

- 20 min to position and characterize the existing offer, 15 min to position and characterize the competitors,
- 45 min to select promising ideas, generate new ones, cluster them and identify/describe/profile a draft concept.

3.3   Concept Description Form for S.PSS and DE

Aims

The objective of the *Concepts Description Form for S.PSS and DE* (Fig. 12) is to finalize the description and characterization of a new S.PSS and DE concept.

Components

It consists of a sum-up of the concept with:

- Concept title
- Satisfaction unit
- Concept description
- Concept profiling, i.e. Provider, Customer, Type of S.PSS, offered Products & owner, Offered services & provider, What is paid, Offer configuration.

S.PSS concept description form

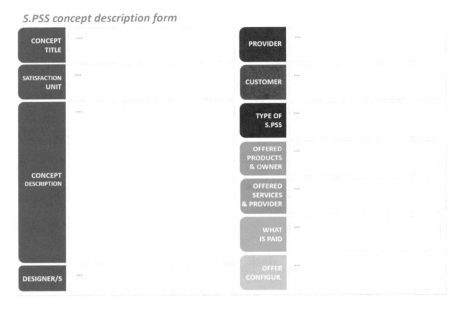

Fig. 12   Concept Description Form for S.PSS and DE

Integrating the tool into the MSDS design process

The Concept Description Form for S.PSS and DE is used in System Concept Design to describe and profile the designed S.PSS and DE concept (Fig. 13).

How to use the S.PSS and DE concept description form

The Concept Description Form can be used in three simple steps (Fig. 12). First, write the title and the description of the S.PSS and DE concept. Secondly, indicate the UNIT OF SATISFACTION of the concept. Finally, characterize the concept with the information in all the fields.

Availability and requested resources

Like the previously described tools, the *Concept Description Form for S.PSS and DE* is an open access tool that can be downloaded for free from www.lens-internationa l.org, 'Tools' section. A computer, a PowerPoint reader, and Internet connection are required to access this tool. The tool may be used by a single designer, though the support of a multi-disciplinary team is preferable. This tool requires at least 15 min to complete.

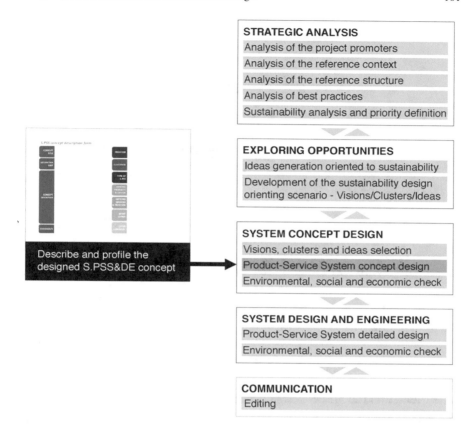

Fig. 13 Concept Description Form for S.PSS and DE integrated into the MSDS design process

3.4 System Map for S.PSS and DE

Aims

The purpose of the System Map for S.PSS and DE (Fig. 14) is to support (co-)designing, visualization and configuration of the system structure, indicating the actors involved and their interactions in distributed systems providing additional support to its users defining DE configuration.[10]

Components

The System Map for S.PSS and DE contains graphical representations of:

- Stakeholders involved;
- Flows/interactions: physical, financial, informational and labour performance;

---

[10]The original System Map tool is presented in detail in the first LeNS book, Sect. 3.6 [10].

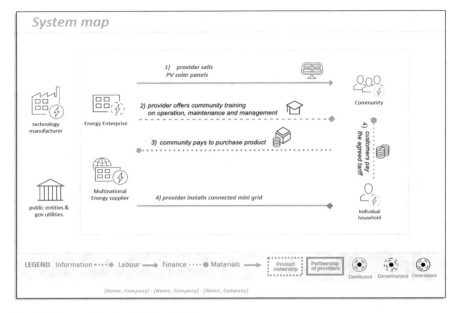

Fig. 14 System Map for S.PSS and DE

- System configurator: Distributed, Decentralized, Centralized.

Integration into the MSDS design process

The System Map for S.PSS and DE is used at various stages of the design process (Fig. 15).

- In Product-Service System Concept Design it can be used to:
  - Visualize stakeholders' interaction within the concept
- In Product-Service System detailed design it can be used to:
  - Further detail and visualize stakeholders' interactions within the concept.

How to use the System Map for S.PSS and DE

The System Map for S.PSS and DE enables comprehensive visualization of the system structure (Fig. 16). To start with, identify boundaries, including offer boundary and system boundary.

Later, identify the actors involved: select a structure icon, then choose a characterization icon to substitute the general one, and finally drag and drop into the system map (Fig. 17).

Now it is time to define interaction flows using arrows and descriptions. Interaction flows can be material flow, information flow, financial flow and labour flow (see the

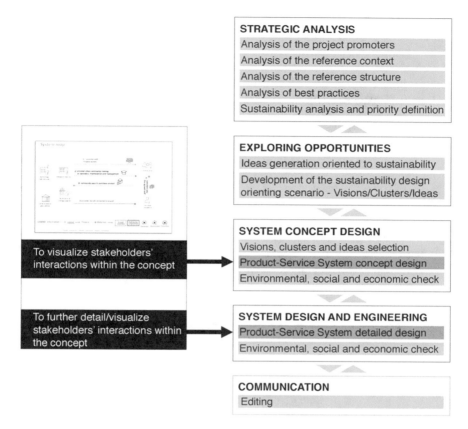

Fig. 15 System Map for S.PSS and DE integration into the MSDS design process

Legend in Fig. 18). Remember that the reading order is essential, thus note the numbering of interaction flows.

Finally, use dashed squares to indicate ownership (owner and product inside) and squares around actors to indicate partnership (Fig. 19).

Availability and requested resources

The System Map for S.PSS and DE can be drawn on paper with no need for software. It is, however, advisable to use slideshow software, in order to facilitate management and modifications. The System Map for S.PSS and DE with labels and icon repositories is open access, available for free download at www.lens-international.org, "Tools" section. The tool is based on a layout and a set of standardized icons, usable with PowerPoint readers. From this base it is possible to modify the various icons and add new ones.

The tool was developed for use by any design team member, and no particular graphic skills are required. The time required to set up a System Map for S.PSS and

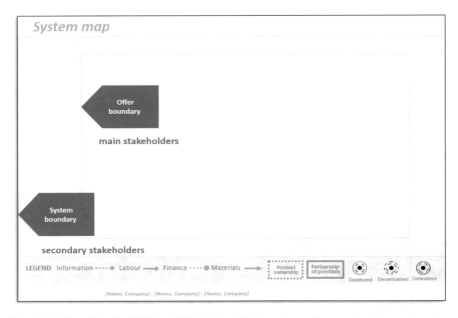

Fig. 16 System Map for S.PSS and DE: design offer boundary and system boundary

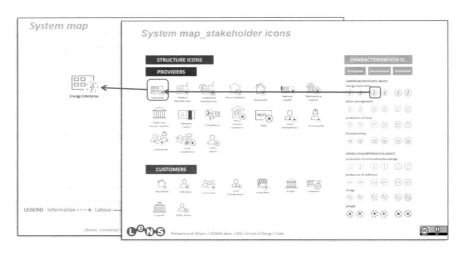

Fig. 17 System Map for S.PSS and DE: design and position actors

DE depends on the level of details along the design process; nevertheless, it could range from approximately 60–90 min.

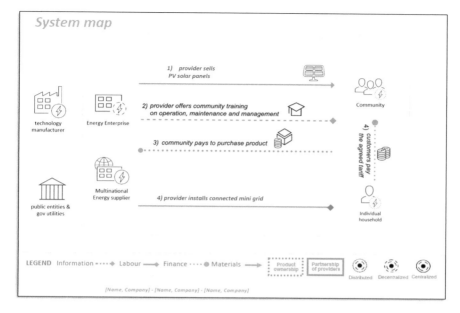

Fig. 18 System Map for S.PSS and DE: design interaction flows

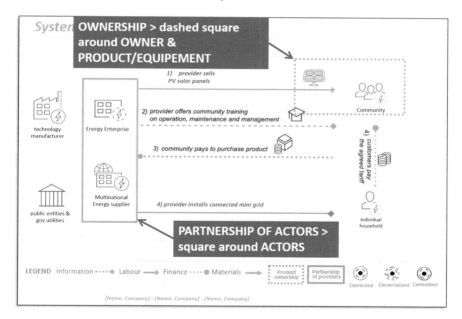

Fig. 19 System Map for S.PSS and DE: design ownership and partnership

3.5 S.PSS and DE Idea Boards (embedded into the SDO toolkit)

Aims

The objective of S.PSS and DE Idea Boards (embedded into the SDO toolkit) (Fig. 20) is to support designers in orientating the system idea generation design process towards sustainable DE for all S.PSS-based solutions.

Components

The tool consists of 6 Idea Boards, one per criteria as listed below, and a corresponding set of guidelines suggesting S.PSS-based DE ideas through innovative stakeholder interactions:

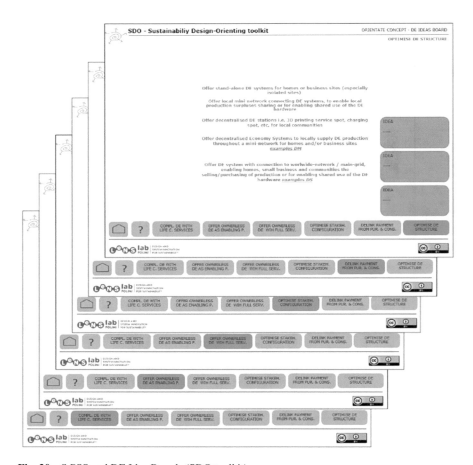

Fig. 20 S.PSS and DE Idea Boards (SDO toolkit)

- Complement the DE hardware offer with Life Cycle services
- Offer ownerless DE systems as enabling platform
- Offer ownerless DE systems with full services
- Optimize stakeholders' configuration
- Delink payment from hardware/resource purchases
- Optimize DE systems structure.

Integration into the MSDS design process

The S.PSS and DE Idea Boards (SDO) are used in idea generation oriented to sustainability to orientate the system idea generation design process towards sustainable S.PSS-based solutions for all (Fig. 21).

How to use S.PSS and DE Idea Boards (SDO)

The following steps must be performed to access the tool:

Fig. 21 S.PSS and DE Idea Boards (SDO) integration into the MSDS design process

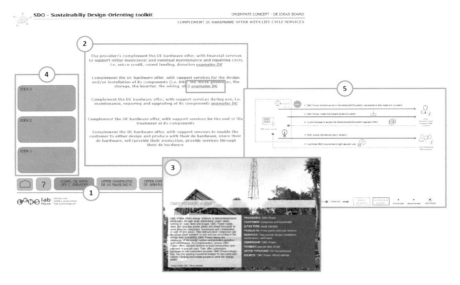

Fig. 22 S.PSS Idea Boards (embedded into SDO toolkit)

- Download the SDO toolkit from www.lens-international.org
- Type project name, etc.
- Click on S.PSS and DE sustainability dimension
- Click on orientate concept.

To orientate the system idea using the S.PSS and DE Idea Boards (Fig. 22), select the idea tables one by one (one for each criterion) (1). Then, read the guidelines (a set for each criterion) (2) and check the guideline's example for further inspiration (3). Drag and drop the "digital post-it" and describe the emerged system ideas (for each criterion) (4). You can see and read more information on the case related to the specific guideline (5).

Availability and requested resources

S.PSS and DE Idea Boards are embedded into the SDO toolkit. The tool is also available for free download at www.lens-international.org, "Tool" page. A computer, a PowerPoint reader and Internet connection are needed to use the tool. Idea Boards require at least 75 min to complete.

3.6 Strategic Analysis Toolkit (SAT) for DE for Socio-Economic Ecosystems (SEE)

The Strategic Analysis Toolkit, SAT, consists of tools which first identify the actors and their activities in the ecosystem; then the infrastructure and needs of the actors; clarifies the goal, problem statement definition, design brief and unit of satisfaction using participatory design tools; and, finally a tool for competitor analysis. This section introduces tools related to processes and sub-processes within the Strategic analysis stage in the MSDS methodology (Fig. 23).

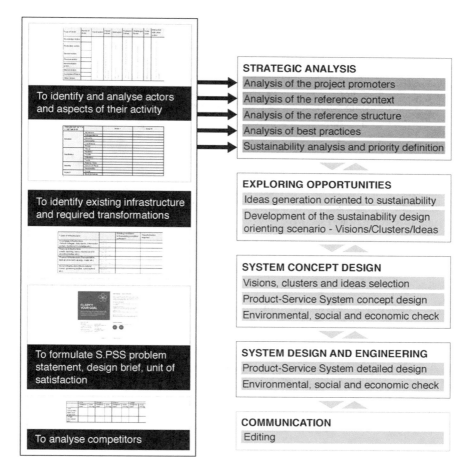

Fig. 23 Strategic analysis toolkit (SAT) integration into the MSDS design process

Aims

The strategic analysis toolkit aims to help a designer in Sustainable Product-Service System Design with an intervention focus on Socio-Economic Ecosystems (SEE) of multi-cultural and diverse communities engaged in distributed economic activities.

Process 1: Project Socio-Economic Ecosystem Analysis

1. *Awesome Actors Tool.* The first step of strategic analysis is to identify all the actors and their aspects of activity, best accomplished by interviewing local administrators and visionaries (e.g. local elders, thought leaders, NGOs, etc.). The Awesome Actors Tool helps its users to identify the main value proposition of the local ecosystem, its problems, all actors and their activities (Table 2).
2. *KFPS Knowledge Mining Tool.* This tool helps to identify existing infrastructure and required transformations. Interviewing local administrators/visionaries helps in acquiring information on service, product-service, and infrastructure transformations planned and required in the local ecosystem (Table 3).
3. *Empathy Mapping,* AEIOU Mapping, Value Opportunity Analysis, SWOT, PESTLE, System Map. A set of tools supports their users in meeting the actual actors and understanding their needs, e.g. Value Opportunity Analysis tool (Table 4).

Process 2: Defining intervention context

4. Co-design using "Clarify Your Goal". The tool adopted from Frog Design [2] helps to define design goals, identify the problem statement, design brief and unit of satisfaction (Fig. 24).
5. Competitor analysis on form, category, generic, budget level (using Porter's five forces analysis if applicable [5]). The tool helps to collect the competition space knowledge (Table 5). Competitors of the system are found based on the clarified goal of the design intervention and the main value proposition of the local context.

Currently, the toolkit has been designed and tested on two SEE contexts, both located in Assam, India.

Availability and resources required

Downloadable files of each tool can be found in Banerjee et al. [1] with the following information on resources and time needed to carry out design processes using each tool.

Table 2 Awesome actors tool

Type of actor	Name of Actor	Contribution	Values added	Motivation	Problems Solved	Challenges faced	Tools used
Knowledge actors		What they are doing?	What value does the actor bring to the ecosystem?	What value does the actor choose to do what it does?	What are the problems that the actor is trying to solve?	What stops it from performing activities?	How do they interact with other actors in the system?
Production actors							
Service actors							
Finance actors							
Administration actors							
Market actors							
Customers/clients							
Other actors							

Table 3 KFPS knowledge mining tool

Types of Infrastructure		Existing conditions: Is the existing condition sufficient?	Transformation required
Knowledge Infrastructure: (School colleges, data banks, information portals, traditional knowledge etc.)	Overall		
	Specific Stakeholder		
Financial Infrastructure (credit, banking, loans, insurance and providing bodies etc.)	Overall		
	Specific Stakeholder		
Physical Infrastructure (Transportation, built environment, energy, water etc.)	Overall		
	Specific Stakeholder		
Social infrastructure (Socio-cultural norms, governing bodies, associations etc.)	Overall		
	Specific Stakeholder		

Table 4 Value opportunity analysis

What are the needs of the actors?		Actor 1	Actor N
Emotion	Adventure		
	Independence		
	Security		
	Sensuality		
	Confidence		
	Power		
Aesthetics	Visual		
	Auditory		
	Tactile		
	Olfactory		
	Taste		
Identity	Point in Time		
	Sense of Place		
	Personality		
Impact	Social		
	Environmental		

Fig. 24 "Clarify Your Goal" section of Frog Collective Action Toolkit

3.7 Distributed Manufacturing (DM) Applied to PSS Design Toolkit

The DM applied to PSS design toolkit (Fig. 25) has been tested with students, experts, manufacturing industry professionals and design practitioners through three rounds of empirical application, to ensure its effectiveness and usability [4].

Aims

The DM applied to PSS design toolkit serves two purposes: (1) it provides its users with knowledge about potential DM opportunities and (2) supports idea generation for PSS solutions improved with DM features.

Components

The toolkit consists of four elements, each of which is described below in detail:

- 40 near-future scenario cards
- 3 scenario cards' selection diagrams
- 1 introductory card
- 1 idea generation diagram.

Table 5 Competitor analysis on form, category, generic and budget levels

	Form		Category		Generic		Budget	
	Competitor name	Value offering	Competitor name	Value offering	Competitor name	Value offering	Competitor name	Value offering
Local-Ecosystem's main value proposition								
Design intervention goal								

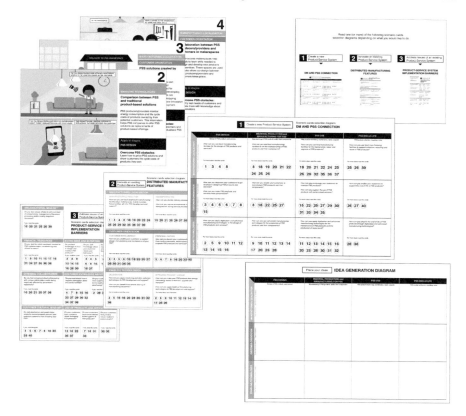

Fig. 25 The DM applied to PSS design toolkit

Near-future scenario cards

Double-sided near-future scenario cards are brief snapshots illustrating how specific features of Distributed Manufacturing can be applied to Product-Service Systems throughout their life cycle (Fig. 26). Scenario cards are made to inspire and to encourage future-oriented thinking. Furthermore, they serve an educational purpose and contain a sufficient amount of information to support a learning process about DM and PSS.

Scenario cards' selection diagrams

Scenario cards' selection diagrams on which scenario cards are mapped illustrate areas tackled by the near-future scenarios. The toolkit contains three scenario cards' selection diagrams (Fig. 27): [1] the stage-by-stage DM and PSS connection diagram (1); [2] the DM features diagram (2); and [3] the PSS implementation barriers diagram (3). These diagrams are made to facilitate relevant scenario cards' selection. Each diagram classifies scenario cards according to PSS life cycle stages and/or DM features, or PSS implementation barriers and contains questions helping to select relevant cards.

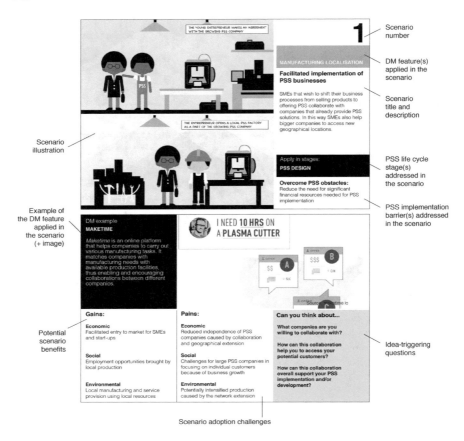

Fig. 26   Front and back sides of the near-future scenario card

Introductory card

This toolkit is made to facilitate new PSS development as well as to improve existing PSS solutions. The introductory card allows the toolkit's users to decide whether they would like to create a new PSS or to improve an existing one (Fig. 28). Depending on their choice, one of the three scenario cards' selection diagrams must be selected.

Idea Generation Diagram

The Idea generation diagram (Fig. 29) is used for positioning ideas developed using near-future scenario cards.

Integration into the MSDS design process

The DM applied to PSS design toolkit can be best used to facilitate idea generation for S.PSS solutions enabled by DM. In addition, near-future scenario cards can be used to explore and analyse existing examples of DM and learn about the DM

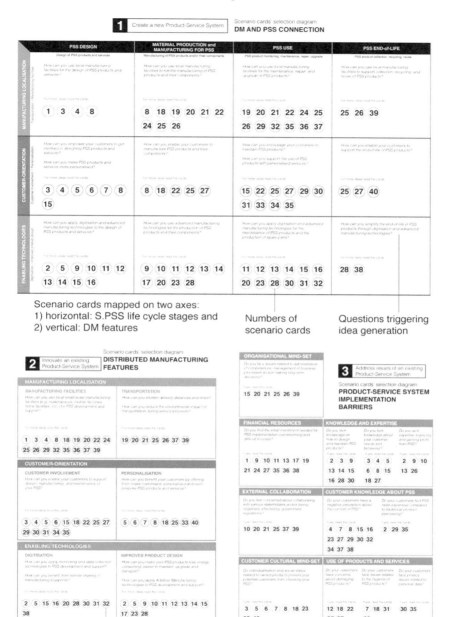

Fig. 27 Scenario cards' selection diagrams

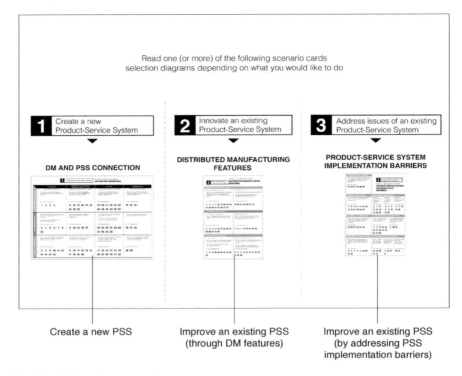

Fig. 28 The introductory card

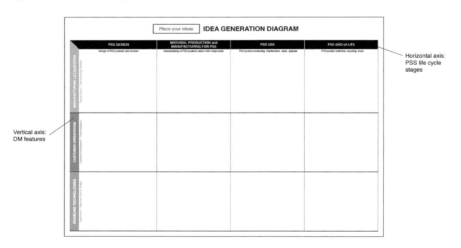

Fig. 29 Idea generation diagram

potential. Finally, the idea generation diagram can be used to position, cluster and select promising developed ideas for further detailing (Fig. 30).

How to use the DM applied to PSS design toolkit

Each element of the DM applied to PSS design toolkit is created to be used in a purposeful order (Fig. 31): first, the identification of the goal using the introductory card (1); second, the selection of relevant scenario cards using the scenario cards' selection diagrams (2); third, DM applied to PSS idea generation using near-future scenario cards (3); and, finally, positioning of developed ideas on the idea generation diagram (4).

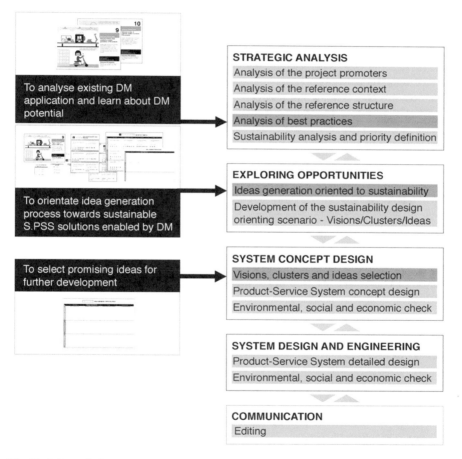

Fig. 30 DM applied to PSS design toolkit's integration into the MSDS design process

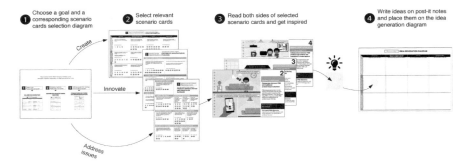

Fig. 31 A proposed design process of the DM applied to PSS design toolkit

Availability and resources required

The DM applied to PSS design toolkit is available for free download (from www.lens-international.org, "Tools" section). The toolkit needs to be printed; other required resources are post-it notes and pens.

The toolkit may be used by a team of designers, design students, or multidisciplinary team. It is advisable to involve various system actors. The toolkit requires at least 120 min to conduct a complete ideation process.

3.8   Summary

This chapter has presented several tools supporting the design of S.PSS applied to DE that have been developed or updated during the LeNSin project. Many other tools have been developed to support system design for sustainability for all and a wide selection of those can be found and downloaded from the LeNS platform (www.lens-international.org).

References

1. Banerjee S, Upadhyay P, Punekar RM (2019) Contextualising sustainable product-service design methods for distributed economies of India. In: Ambrosio M, Vezzoli C (eds) Designing sustainability for all: Proceedings of the 3rd LeNS world distributed conference, vol 1, pp 270–275
2. Frog Design (2016) Collective action toolkit. https://www.frogdesign.com/wp-content/uploads/2016/03/CAT_2.0_English.pdf
3. Manzini E, Collina L, Evans S (eds) (2004) Solution-oriented partnership. Cranfield University, Cranfield
4. Petrulaityte A, Ceschin F, Pei E, Harrison D (2020) Applying distributed manufacturing to product-service system design: a set of near-future scenarios and a design tool. Sustainability 12(12):4918

5. Porter ME, Kramer MR (2006) Strategy and society: the link between competitive advantage and corporate social responsibility. Harvard Bus Rev
6. Tischner U, Vezzoli C (2009) Product-service systems: tools and cases. In: Crul M, Diehl JC (eds) Design for sustainability (D4S): a step-by-step approach (United Nations Environment Programme UNEP)
7. Tukker A, Tischner U (eds) (2006) New business for old Europe: product services, sustainability and competitiveness. Greenleaf Publishing, Sheffield, UK
8. van Halen C, Vezzoli C, Wimmer R (eds) (2005) Methodology for product service system. How to develop clean, clever and competitive strategies in companies. Van Gorcum, Assen, Netherlands
9. Vergragt PJ (2002) Strategies towards the sustainable household. Final report. TBM, Delft University of Technology, Delft
10. Vezzoli C, Kohtala C, Srinivasan A, Diehl JC, Fusakul SM, Xin L, Sateesh D (2014) Product-service system design for sustainability. Greenleaf Publishing, Sheffield, UK

S.PSS and DE in Practice

Brenda Garcia Parra, Cindy Kohtala, Tatu Marttila, Aguinaldo dos Santos, Sandra Molina Mata, Fang Zhong, Nan Xia, Xin Liu, Jun Zhang, Sharmistha Banerjee, Pankaj Upadhyay, and Ravi Mokashi Punekar

1 Introduction: Teaching and Learning Contemporary Design for Sustainability

Contemporary challenges related to sustainability are shared across the globe. Their materializations, prioritizations and emphases, however, vary from one region and context to another. As we have seen in the previous chapters, Sustainable Product-Service Systems (S.PSS) and Distributed Economies (DE) as concepts are still in the making, and tools to assess and implement them in design are still developing. Their interpretations can also take various forms when they become introduced into different contexts. Alongside these shared challenges, there are also specific regional or historical tensions, which connect not only with education, design and the histories and trajectories of industrialization, but also arise in international projects and collaboration. Such tensions become even more evident if new concepts and contents come into play. It can also lead to differing interpretations on how to approach them.

B. Garcia Parra (✉) · S. Molina Mata
Universidad Autónoma Metropolitana, Mexico City, Mexico
e-mail: bgarcia@cua.uam.mx

C. Kohtala · T. Marttila
Department of Design, Aalto University School of Arts, Design and Architecture, Espoo, Finland

A. dos Santos
Universidade da Região de Joinville (Univille), Joinville, Brazil

F. Zhong · N. Xia · X. Liu
Tsinghua University, Beijing, China

J. Zhang
Hunan University, Changsha, China

S. Banerjee · P. Upadhyay · R. M. Punekar
Indian Institute of Technology Guwahati, Guwahati, India

© The Author(s) 2021
C. Vezzoli et al. (eds.), *Designing Sustainability for All*, Lecture Notes in Mechanical Engineering, https://doi.org/10.1007/978-3-030-66300-1_5

The international LeNS network was created in an EU-funded project between 2007-2010, and it has continued to expand and interact in the LeNSes project (2011-2015, with a focus on renewable energy for all) and in the LeNSin project (2015-2019). With its 'ethos' on 'multi-polar' collaboration and several strong regional networks, the international LeNS network has accommodated an atmosphere that has been sufficiently open and sensitive to elaborate these concepts further, and to critically assess their potential in developing new, more sustainable solutions.

This chapter shares the experiences from the LeNSin seminars and pilot courses from various perspectives. In this chapter, we also briefly discuss the potential of design education as a transdisciplinary matchmaker between various actors and networks.

2 Introducing S.PSS and DE into Higher Education in Design

Despite its recent origins in the mid- to late twentieth century, discourses around professional design, its identity and expectations, are relatively consistent across the globe, particularly with regard to industrial design. Various methods for sustainable design, including product-service system design, have also evolved, increasingly acknowledged in public discourse, such as in EU and UNEP publications. However, the very concept of PSS still allows for various interpretations depending on the socio-economic context. For example, in some pedagogical contexts, the concept of a 'sharing economy' may be seen as more engaging or understandable for design students than 'user-oriented PSS'. Moreover, when relatively broad concepts such as Distributed Economies are introduced into sustainable design teaching, interpretations can vary significantly with regard to focus and expectations.

Design activities have gradually extended further from the studio and the factory line. Currently, design connects with various domains of interest, with products and services, but also systems innovation; with organizations and business, but also societal change-making. Designers work with diverse professionals and experts, as well as laypeople and public media. This diversity extends the area where such interpretations can be trialled.

In this section, we reflect on how S.PSS and DE as concepts can be introduced into different geographical, socio-cultural and educational contexts, and we examine some of the choices and emphases in developing two rounds of pilot courses during the project. We also address the variety in which the DE focus can be adjusted and look into the role of the university in contemporary knowledge building for transdisciplinary sustainability.

One integral aspect in the LeNSin project has been in sharing experiences on teaching and in developing new educational content. The main strength has been the strong network, which has helped to overcome practical difficulties, and to balance course expectations and institutional constraints in developing new teaching contents.

2.1 Experiences from Regional Pilot Courses: An Overview

Sustainable design has gradually become a highly promoted strategy linking industrial developments, consumer domain actions and policymaking. In sustainable design, as often in complex problem-solving processes, several actors from different fields need to work towards a shared goal, and a more detailed discussion on the driving values and goals pursued is needed. The challenge of sustainability lies in connecting not only scientific research and politics, but also the perceptions and actions of professionals and laypeople. In this sense, design for sustainability can be understood by its very nature also as transdisciplinary, drawing together considerations from ecological, societal and economic domains into a shared process of mediation and making.

Contemporary design activities in various regions are in many ways still based on the educational programme of the Bauhaus school, where architects, painters, and sculptors combined multiple perspectives with an emphasis on workshop or studio work [4]. In the last few decades, however, the role of design has gradually shifted towards higher levels of focus, from the crafts studio and factory line towards society at large and towards broader socio-technical systems [6]. Today, design has been noted as a possible catalyst for social innovation [10] and sustainability transitions [9]. Consequently, the potential of design has been increasingly noted also in relation to transdisciplinary activities in education and in sustainability at large ([5, 8, 14]).

Design schools around the world share similarities in both the challenges they face, as well as the potential that the discipline itself allows. As a discipline in higher education, design often connects with engineering and business, but also media and art. Earlier, it has acted as a bridge between the producer and consumer world. Particularly now, with new phenomena heralding new agency and competencies for users and consumers—as seen in the spread and promotion of social innovation, the influence of the internet and peer networks, and the DIY, amateur design 'maker movement'—these roles have also become increasingly mixed [11].

2.1.1 Developing Teaching in Two LeNSin Pilot Rounds

The LeNSin project focused on developing new teaching contents on S.PSS and DE, but also on expanding the network of partners, to gather an understanding of various DE related actors in different contexts and countries. This was taken forward in the form of case studies, new tools and methods, local seminars to gather insight, and in consecutive pilot courses in which various DE topics were taken under study with students (DE topics are discussed in Chap. 2). The seminars gathered local actors that shared an interest in the topics, but also linked to already existing networks and projects. The consecutive pilot collaborations took place with design students from various fields, ranging from media and graphic design to industrial and service design, and to engineering and architecture.

To understand what impact the project has had, we look into the interactions in developing these collaborations and reflect on the preparation process, the emphases taken in the actual pilot courses, as well as their outcomes. Our reflections are grounded on the course materials (syllabuses, teachers' course reports), our own experiences and insights from interviews with the teachers involved.

Overall, five seminars and ten pilot courses were organized in five countries, with two main partners from each country and additional associate partners around the region. This interaction also constituted one main part of the whole project, where theoretical contents, design methods practice and real case studies came together. Initially, in each country the two partners came together to host a seminar, in which the main topics of S.PSS and DE were discussed from the regional perspective. Later the pilot courses were conducted, which aimed at examining and designing for DE in the various contexts in student case work. Teachers refined their understanding of the connection between S.PSS and DE during the courses, as well as tested and refined design tools. In China, the focus in the first pilot course was on lighting and 3D printing, in both commercial contexts and marginalized communities, and in the second pilot on regional food culture. In India, the focus of the first pilot was to help a local NGO actor boosting regional health, well-being and resilience and the second was on developing the local silk weaving industry. In South Africa, the focus was first on developing distributed health solutions and information and then on developing a supporting app for deaf people. In Brazil, the case work in the first pilot focused on the local fashion cluster and the second one on local mobility. And lastly, in Mexico, the first pilot focused on a local book club programme and then the second pilot on the university payment service system for students. Finally, besides access to a gradually extending case study library and improved revisions of toolsets, student teams also had an opportunity to submit their solution to the LeNSin student competition, and eventually six national winners were selected (including Europe), and four honourable mentions were given [12].

Overall, the themes in the regional educational activities progressed in different directions. However, the predefined structure of interaction and thematic content helped to keep a relatively coherent whole. Regional seminars and the following pilot courses called for a cumulative amount of preparation, but also ensured that the network of actors was gradually formalizing and roles became clear. In the end, the teaching activities involved an extended number of educators, both from the local region as well as internationally, and they attracted local attention.

During the teaching collaborations, in many locations, there were also unexpected local events—natural or societal—that led to additional challenges in preparations (for example, employee union strikes, student strikes, political instability and natural disasters such as earthquakes). The strong network allowed the necessary reflexivity that helped to overcome these obstacles.

Although each pilot comprised introduction to S.PSS theory and tools and an introduction to DE topics, the structure of the pilots varied. As the participating teachers visited several locations, and experiences were shared across, the teaching as a whole nevertheless remained rather coherent in relation to its main topics. Additionally, various experiences with tools and methods for teaching were exchanged during the pilots, but also informally across the network via email conversations or face-to-face meetings.

The structure of the seminars and two consecutive pilot courses held in each region provided the possibility to have 'rounds of iteration'. Each pilot also had visiting teaching partners from another university, as well as observers from a third one. Experiences were then gathered in reports and exchanged in project meetings. This material also allows for subsequent academic communications and reflections in various forms.

Although it was challenging to insert an intensive, short course within most of the institutions involved (see Sect. 6), many of the teachers later reported that the very intensity was beneficial to the students' learning. Many things about the pilot course were 'new' for both students and teachers: the diversity of didactic approaches (from theoretical lectures to active teamwork and fieldwork); the diversity of the student body (e.g. coming from all over the country, see Sect. 5, or from different departments, see Sects. 3 and 4); and the diversity of perspectives represented (teachers from other countries, stakeholders from companies or NGOs, and so on). Teachers quickly learned to improvise and take advantage of each other's expertise, while needing to create a learning structure that did not lose students through the gaps. Teachers later appreciated how these opportunities and challenges helped create courses that managed to avoid "superficial sustainability" or "sustainability-as-usual", as a kind of green paint splashed onto design education. Students were rather pushed to improve their abilities in systems thinking and to imagine and aim for new paradigms beyond business-as-usual: the territory of Distributed Economies where locally relevant solutions with greater sustainability potential are identified and fostered or designed anew and gradually embedded within the existing culture. S.PSS and DE were unquestionably often problematic concepts, but both teachers and students worked on translating the terms literally and culturally: reframing, re-coding and re-interpreting them. In some cases, the internationality of the work helped to raise the profile of sustainable design education in the institution and lend it further legitimacy, in a global context of tight budgets and instrumentalist learning objectives.

2.2 Reflections: Teaching S.PSS and DE Design

The concept of S.PSS is rather established in both design teaching and industry in many regions, and as a concept, it also acts as a suitable basis to develop a new understanding on DE. DE as a thematic area of focus, however, introduces very different interpretations in different contexts, regarding expectations, mode of work and developed outcomes. One important outcome is, in this sense, also in being able to discuss these views and to spread it forward to new actors. Getting to grips with what Distributed Economies actually means and why it is a beneficial umbrella concept requires much discussion among teachers and students on what kind of industrialized or post-industrial context they exist within and how it compares to others. It is pedagogically useful to make the concept familiar, to bring it 'home', by identifying local case studies that can be classified as various DE cases, whether distributed manufacturing or distributed renewable energy. This, in turn, helps identify the case's sustainability benefits and threats, as a locally relevant system with cultural, social, technical and economic aspects. Teaching and learning DE is therefore not a case of importing a European concept into a non-European socio-technical environment, nor is the intent to design a solution that imitates solutions from the global North. Instead, what is important is to define 'sustainability' in dialogue and according to what is locally appropriate.

Adams et al. [1] promote developing education based on a "sustainability culture conceptual framework", which connects people, teachers to other staff to students to external stakeholders, and that entails organizational transformation: building systems that support dialogues on both visible artefacts and activities and invisible values. Consequently, when we introduce design collaboration into the context of sustainability, its driving values are challenged, and responsibility and ethics come into play. To overcome these obstacles, collaboration is needed across continents and disciplinary sectors. In this process, projects as arenas to facilitate these discussions have high impact—and an open and supportive network helps.

Sustainability and 'sustainable development', in the end, are wedded to (global) equality, equity and justice, roles, access to participation and transparency. To this end, if design practitioners have a role in promoting collaborative mediation for sustainability or even further—to promote democratic assessment of heterogeneous perspectives for sustainable innovation [13]—this also calls for fundamental changes in how to approach design education and its processes of teaching and learning.

And yet, design activities around the world are fundamentally grounded on iterative development. Design thinking acts in bridging problem and solution spaces [7], and its activities proceed by default through trial and error. Design as a discipline remains a developing field, continually producing new methods and collaborations in various contexts, in between and in connection to multiple domains and discourses. And finally, at best, teaching design involves an open and expansive process. Contemporary design activities involve several emphases on inducing and promoting collaboration and shared mediation. Collaborative, participatory design processes can support shared knowledge building and development of practice. Such interaction can also connect with local and tacit understanding, to be adapted and better applied in new contexts.

2.2.1 Discussion: The Changing Role of (Design) Academia

When design educators are networking globally and bringing local actors into dialogues to promote sustainability in various contexts, conventional industrial collaborations can expand further into new networks (see Sect. 3). S.PSS and DE as concepts allow such expansion and extend these networks further.

In developing new international collaboration on teaching and making, interaction needs to be embedded in a shared and reflexive process. In support of this, design remains an open field for education and action, linking various local and global inquiries across several professional domains. And as a result, design for sustainability as an aim and agenda can support a transformation in contemporary practices of making and learning; design acts as one key focus for developing policies and action, attracting interest in developing new ideas for societal sense-making.

Today, universities are adopting a new role, to establish their position in the political and economic structures of an increasingly knowledge-driven society. This new role emphasizes knowledge production for society and societal benefit, calling for stronger connections between research, education and everyday practices to expand participation to the outside world. For contemporary universities, this call moves the emphasis on how students and other stakeholders in the processes of learning are taken into account when joining up the fundamental orientations for any action.

As a mode of interaction and collaboration—and shared development of learning content—the LeNSin pilot course interactions provided a valuable opportunity to develop new tools and methods to implement sustainable design, and to share and connect the topics further. In parallel with the pilot courses, other curricular courses and collaborations with stakeholders (NGOs, municipal authorities, companies and so on) furthered the lessons learned. In the following sections, we will describe further how collaboration particularly with external stakeholders in the courses is carried out, from fieldwork involving regional industry clusters to small NGO partners in a long-term partnership in education.

3   Working with a Regional Industry Cluster in Education in Brazil

In the north-eastern part of Brazilian territory, nearly 23.5 million people have faced harsh living conditions due to severe weather for decades. This so-called Semi-Arid region is characterized by high temperatures and low rainfall, generating water shortages. Poverty and social injustice have emerged as problems associated with the scarcity of water, but policy and top-down solutions have failed to democratize access to the water supply, hindering local development and putting individuals under economic, political and cultural domination. However, in recent years, with the implementation of social innovation initiatives, local communities of the Semi-Arid region have begun to change their dependence on centralized public policies and to develop bottom-up alternatives to mitigate the conditions associated with water scarcity.

Based on innovative interactions between stakeholders, the Agreste's fashion cluster was created. The cluster specializes in the manufacturing of jeans, cotton and polyester clothing. A total of 18 000 SMEs employ 8% of the workforce in the state of Pernambuco and generate 5% of the state's GDP. Despite all the economic benefits brought to the Semi-Arid area by the fashion cluster, the region's environmental and social degradation has worsened. It was impacted by the intense use of already scarce resources, such as wood and water, which are used in the manufacturing process, mainly for the washing and finishing of jeans. From 60 to 100 litres of water are necessary to wash one pair of jeans. Toritama, the city where jeans are produced, manufactures 800 000 pairs of jeans per month. The waste from this process is not correctly processed. It is disposed of in the river that crosses the city, changing the colour of water from pink to blue to a coloured mix.

3.1 The LeNSin Pilot Course Brazil

In June 2017, the first LeNSin Pilot Course Brazil took place in the city of Recife, at the Federal University of Pernambuco. Thirty-five students from the business administration and design courses from UFPE were challenged to develop S.PSS proposals for the fashion cluster of Pernambuco. The students were expected to develop an understanding of current environmental issues; demonstrate understanding of the tools used to develop S.PSS concepts; discover design strategies and to design an S.PSS for Distributed Design and Distributed Manufacturing with a particular focus on the Brazilian context; and to explore and test out-of-the-box concepts and ideas for S.PSS concepts. A field trip was organized to the fashion district to collect data and to give participants the opportunity to 'experience' some of the issues described in the challenge. Students also presented and validated initial ideas with representatives of the fashion cluster. Their final design concepts ranged from ways to transform waste from manufacturing processes to solutions to promote the empowerment of women in the region.

From the point of view of learning objectives, the mixture of students from different disciplines was beneficial: it proved to be quite effective for the learning process to have business and design students together in mixed groups. Having the participation of lecturers from other universities and presenting local and international case studies brought together the local and the global perspectives on S.PSS, fostering a better understanding of the concepts and enriching the discussions. That said, sustainability is complex and much information was presented, needing systematic and constant reviews throughout the course.

The collaboration with the regional industry cluster was important for several reasons. First, there is an urgent need to put our students in contact with the context in which they live. For instance, many of the students did not know the possible negative environmental impacts of the clothing they were wearing. Coming into direct contact with the problem through a site visit enhanced the creative process and resulted in a more empathic process, even if such visits are time-consuming within short courses. Furthermore, the field study sped up the bonding among team members, resulting in a pleasant atmosphere for the practical part of the course. The second reason the collaboration with industry was essential is because it shows potential partners outside the university, the role academia can play in practice, with concrete tools and innovative ways for understanding their problems.

3.2   Reflecting on Industry-Academic Collaborations in Brazil

There are at least three key challenges for setting up a direct collaboration with industry, based on the Brazilian experience. The first relates to information scarcity. Achieving a robust (meta) concept on PSS and DE requires a wide set of data, information and intelligence that is not usually ready to be used by students. The inherent nature of Distributed Economies often implies the consideration of stakeholders that conventional companies have not integrated into their business process and, therefore, have little knowledge as to how to support the creative process. In other cases, the business partner cannot disclose information as it often deals with strategic and sensitive issues, such as the long-term vision and objectives regarding the service and product portfolio. In order to enable a meaningful experience, the approach adopted in the pilot courses in Brazil was to present a compact set of information about the problem, leaving some room for the students to collect additional information as needed. Although students can opt to adopt more empathic approaches with the stakeholders (such as focus groups) or more quantitative approaches (such as Business Analytics), short courses have shown that it is more viable to dedicate time to the analytical process than the data collection. Awareness of the scope and depth of the information required by S.PSS applied to DE Design is, therefore, one of the expected learning outcomes of these courses.

The second challenge relates to expectations regarding innovation insights. Expectation management is quite important when developing courses on S.PSS applied to DE. When the business partner is not fully aware of the meaning of PSS and DE, the expectations may be overoptimistic regarding what would be delivered at the end of the course. While on a regular product design course a student may be able to produce a usable prototype, tested in a real-world setting, the complexity of an S.PSS applied to DE problem usually allows the students to only get to the (meta) concept stage. When the business partner is knowledgeable about S.PSS and DE the expectations are naturally more realistic. In such situations, the (meta) concept produced by the students results in insights for the business partner, which is the most relevant benefit of the cooperation. No ethical issues have been raised in any of the pilot courses in Brazil in this academia-industry collaboration, since none of the student groups have reached a stage where an idea could effectively result in e.g. a patent. Indeed, most of the projects developed by the students have achieved more innovation at the system and service level than at the product level, making it difficult to reach a stage where copyrights would be an issue to be raised.

The third key challenge in industry-academia collaboration is being part of the learning process and not just a client. Comparing the experiences in Brazil, it is quite clear that a full involvement of the business partner contributes to a better result with regard to the learning process. Such involvement might require a wide set of contributions: giving technical and managerial feedback on the evolving concepts; reassuring the students regarding the attractiveness (or not) of their concepts; pointing out barriers and strategic advantages of their ideas in topics that have been overlooked by the students; bringing onto the table insights from past experiences; information about the dynamics of the stakeholders in the industry, and so on.

When the industry is already involved and interested in PSS projects, the motivation to take part in the learning process of the students is two-fold: to have direct contact with methods and tools developed in academia, and to contribute to the training of possible future employees. In Brazil, there is a growing demand for design professionals with competencies in PSS and Service Design. Actively developing new young professionals, observing them in action, offers the partner companies the opportunity to identify new talents that might be recruited to join their staff.

4 Working with a Regional Industry Cluster in Education in India

This section presents the experience of the team at IIT Guwahati in teaching Design undergraduate, postgraduate and Ph.D. students the principles of Design for Sustainability (DfS) for Socio-Economic Ecosystems (SEE) of India [2, 3]. According to Banerjee et al. [2], "A SEE is a context where the economic activities of the community are deeply ingrained in the socio-cultural ways of living." In these contexts, the major challenge for DfS is how, through design, one can bring about:

- *first, the sustainability orientation to the socio-ethical dimension in a manner that it is in the economic interest of the system stakeholders to be so, and;*
- *then, the sustainability orientation to the environmental dimension in a manner that it is in the economic interest of the system stakeholders.*

Another characteristic is that it is difficult to identify one company or stakeholder who is the promoter or provider of the offerings of the SEE. Instead, these are multi-stakeholder ecosystems. The inherent nature of the economic activities in these contexts is distributed (Distributed Economies) in nature. SEE might be distributed in terms of design, manufacturing and knowledge generation. These ecosystems might also have a long history of existence and, as a result, have evolved their system to be sustainable on many accounts. In order to initiate any design intervention, a designer must therefore deeply study these traditional ecological and social knowledge systems and their integration with the local cultures.

4.1 Case Study Location: Sualkuchi Silk Handloom Industry as a SEE

Sualkuchi in the Kamrup district of Assam, India is a census town and is made up of a cluster of 16 villages. It is on the banks of the river Brahmaputra, 35 km from Guwahati, the largest city in the northeast of India. The population is more than 100 000. It is also famously referred to as the 'Manchester of Assam' due to its large silk handloom weaving industry which now also has a trademark—Sualkuchi's. The handloom industry here is even mentioned in the works of Kautilya, an Indian royal advisor and economist who lived during 371–283 BC. The current form of the industry is a result of the encouragement it received during the Ahom Dynasty from 1228–1828 AD [15].

A typical household in Sualkuchi owns at least one loom and contributes to the silk weaving industry here. Post-independence of India, the industry began to flourish and reached its peak during 1981–2001 when looms per household increased from 2 to 6, on average [17]. During this time, many households shifted their operations towards entrepreneurship, owning 50 or more looms, employing weavers rather than using the family members as weavers. There are four major categories of actors in the ecosystem: owners, weavers, reelers and helpers. The owners might be small (<5 looms) or large (>50 looms) and own the instrument of production, the Jacquard loom. The small owners mostly weave and reel themselves with their family while others hire weavers, reelers and helpers. The contracted weavers are paid based on the length of garment woven and the number of design elements. They learn to weave on the job and come from all over Assam. Some of them stay back in Sualkuchi while others go back to their native place to start their handloom setup. The reelers are also contractual and perform pre-loom activities like reeling and spinning of yarn while the helpers are paid monthly for helping the other three actors. Other standalone actors support the ecosystem: designers, loom makers, and servicers, intermediaries, distributors, shopkeepers (selling raw materials, selling finished products), government support units for low-cost raw material for small owners, silk testing lab, and Sualkuchi Tat Silpa Unnayan Samity. The biggest strength of the existing system is its distributed nature in terms of design and manufacturing (it has very few large units). Attention to technology, design and business model upgrading has lacked due to unorganized production systems, leading to stagnation. There are also rising costs of raw materials and lack of a financial support system, meaning the small owners are slowly disappearing leading to possible centralized economic models kicking in.

The primary learning objectives for the course were "Developing competencies", "Creating and changing values, attitudes, and awareness", "Transferring knowledge and understanding", "Promoting sustainable behaviour and responsible action" and "More just and sustainable society". Students were introduced to the history, development, approaches and various tools for DfS in a global context, as well as how to tackle DfS for SEE in the Indian context. Through field study methods, they were encouraged to identify how indigenous systems have evolved to live in harmony and a mutually symbiotic relationship. This also entailed identifying what new challenges were entering the system and how they are challenging the sustainability (social, environmental and economic) of the system. Given this background, the students would then design for the emerging context using the fundamentals of DfS, SEE and Distributed Economies. Lectures were organized by local stakeholders, visionaries and administrators, along with faculty and researchers from Design, Engineering and Social Sciences. The students also came from diverse backgrounds, design, architecture and fashion.

In the first process, the 'Project Socio-Economic Ecosystem Analysis', the group of actors from the ecosystem were identified who will together own the new S.PSS and their critical activities, by interviewing the local administrators and visionaries. They can quickly provide the designer with the main value proposition of the local ecosystem, its problems and an understanding of all the actors and their activities. The interviews also provided valuable information to help identify the challenges, potential barriers and support for the S.PSS to be designed in terms of infrastructure (knowledge, economic, physical or social) and changes required. Using mapping tools, the students could then identify the needs of the actual actors.

In the second process, 'Defining Intervention Context', the context for intervention was identified using a participatory approach, involving as many actors of the SEE as possible. This resulted in the identification of an S.PSS problem statement, design brief and unit of satisfaction. In the light of the selected problem statement, the students could then conduct a competitive analysis on two ecosystem parameters: the local ecosystem's main value proposition and the design intervention goal.

4.2 The Outcomes of the Course

The main outcome of the course was a shared Living Lab in the SEE for constant design upgrading and archiving in collaboration with the local NGOs, Government, entrepreneurs, educational institutes, designers, machinery and software manufacturers. This configuration thus reduces the cost of design and keeps design up-to-date with current fashion trends. A design concept for a co-working space was also developed, as well as a central online platform for global customers' orders and customization offers.

The collaboration with local stakeholders was fruitful, as it emphasized to the students that we should not teach and learn sustainability as a criterion in the process of design, but design in the context of sustainability. Validation of ideas with stakeholders is vital to keep students grounded in the context and for the solutions to be useful. However, students find systems thinking complex and intimidating in the context of sustainability when the cascading impact of one decision can lie in multiple aspects of the system. Repeated one-to-one discussions with the instructors were needed to ensure final design solutions were oriented to the context the students had analysed. Moreover, having a range of faculty members to support was beneficial, as we need integration of multiple knowledge domains. This range can help with the constant tension in teaching between breadth versus depth of analysis and ideation in courses of limited duration.

5 Country-Wide Teaching Networks on Sustainability: LeNS China

In China, two pilot courses (Tsinghua University and Hunan University) were organized involving about 30 teachers from home and abroad, and around 150 students from more than 20 universities across the country (including almost all LeNS China member institutions and other universities offering sustainable design-related courses), as well as practitioners in the field of sustainable design.

5.1 Pilot Courses in China

The courses provided an international communication platform for people to promote and spread sustainable design in China. The teachers systematically combined the relevant knowledge of sustainable design for the students—history, basic concepts, methods and tools—as it is important for all students to discuss and think according to a common understanding and a unified paradigm. In the courses, the teacher teams not only encouraged students to think comprehensively about sustainable design from the environmental, social and economic levels, but also guided students to integrate 'culture' as an element into the design. In addition, students were encouraged to build their own understanding of sustainability and explore innovative and sustainable solutions. To address the concepts of S.PSS applied to DE (Distributed Manufacturing and Distributed Renewable Energy, see Chap. 3) the students were taken on field visits to relevant cooperating companies, on the one hand, to understand the most cutting-edge technology development and applications, and on the other hand to encourage students to consider sustainable solutions in the future from a commercial perspective. Sustainable solutions should not only exist at the concept stage, as they require effective technical support and reasonable commercial promotion to realize fully.

Organizing such a large pilot course was both compelling and challenging. It was compelling that students showed a strong interest in the background: the international cooperation of the teaching team and the subject. With different cultural backgrounds, academic backgrounds, novel ideas and a vision for a sustainable future, these future designers were coming together to communicate sustainable design ideas and share sustainable design practices, experiences and insights, which was not only beneficial to students, it was also for teachers and all participants. The difficulty was due to a large number of students: the degree of sustainable design knowledge and understanding was uneven, and it was, therefore, challenging to conduct more in-depth discussions during the course and for students to come up with more reflective opinions and ideas. The time pressure of the short, intensive courses nevertheless stimulated students and pushed them to their full potential. Behind each final presentation was the discussion, debate, disagreement and compromise of the students, which represents the meaning of teamwork. Despite this, it was an extraordinary experience and learning opportunity for the teachers and students who participated.

Tsinghua University Academy of Arts & Design and Hunan University School of Design & Art are among China's top design schools. As the organizer of the LeNSin pilot courses, both have great appeal and influence on other design schools in China. On the one hand, teachers in colleges and universities want to learn about the resources, teaching methods and curriculum materials of sustainable design teaching, and have more exchanges with domestic and foreign counterparts. On the other hand, students hope to master the cutting-edge knowledge of the design field and learn design thinking through the curriculum training.

5.2 Having Impact Nation-Wide

In China, teachers and students are increasingly interested in and becoming more involved in sustainable design, research and discussion. With the push of government policies in the field of sustainable development, many institutions are gradually opening courses related to sustainable design, and students are increasingly willing to reflect sustainable thinking through their projects. Therefore, there is a great need for the study of theories, methods and tools for sustainable design.

Sustainable design is empowered by its cross-disciplinary nature, by inviting not only international teachers, but also provide a global vision, broaden the horizons of students and promote cultural exchanges. Sustainable design is an area of continuous development and evolution, and it is also a process in which teachers and students learn and explore together. Designers must be conscious and responsible for their decisions to ensure that the design is positive for people, as the products and services we create will influence and change people's lives to a large extent. Sustainable design means that we must go beyond traditional design thinking to look at design innovation in a more systematic and integrated perspective. The ultimate goal of sustainable design is to achieve a win-win situation for social benefits, environmental protection and economic development. This course is part of this effort.

LeNS China is China's most active sustainable design teaching alliance. The two pilot courses organized by this project not only effectively promote the communication of sustainable design concepts and the exchange of teaching experience in design institutes, but also have impact more widely. The related course information has been widely disseminated on the WeChat platform and the courseware and lecture materials have been downloaded nearly a thousand times on the LeNS-China network platform.

Sustainable Product-Service Systems (S.PSS) as a design strategy, aimed at exploring how to understand and intervene in these new, emerging economic and social forms, generates opportunities for sustainable business model innovation and industrial value creation. In this process, new tools and design methodologies need to be included to continuously meet the requirements of economic, environmental and social sustainable development. Through curricula, teaching and learning of

S.PSS ideas, and the participants' use of methods and tools, more applications will be generated and new practices conceptualized to adapt to China's conditions.

6 Integrating Experimental Pilots into Long Curricular Courses

This section highlights the importance, challenges and opportunities of executing a pilot course during the complete term of an existing curricular course, which has previously established course objectives, credits and enrolled students in a full-time programme, and sometimes pre-assigned teachers who may (or may not) be familiarized with a Sustainable Design paradigm.

Incorporating a pilot course within a curricular course according to its full-time scheme sets a significant challenge, but also offers the opportunity to have a wider perspective regarding the trajectory of an enrolled student at an undergraduate design programme. It allows teachers to be able to identify what kinds of preliminary knowledge would be needed and when, to recognize what specific topics should be previously introduced or reinforced in strategic courses along the complete undergraduate programme, to promote interrelations with other curricular courses, and to understand how S.PSS and DE knowledge contributes to shaping the overall graduate profile of the students.

6.1 LeNSin Pilot Courses at UAM Universities

The Autonomous Metropolitan University (UAM) participated in the LeNSin Project involving two campuses, Azcapotzalco Campus and Cuajimalpa Campus, both in Mexico City. Though both precincts have different undergraduate programmes (Industrial Design and Design, respectively) they both share a general structure, in which courses are scheduled by Trimesters and envision a Three-Trimester-long course at the end of the undergraduate programme called "Final Project", in which students are to immerse in a complex problem, propose, evaluate and communicate a design solution through a thesis. The Final Project is thereby part of the UAM design programme's strategy to provide students with the time and academic space to have all previously acquired knowledge during the Bachelor's programme put into practice. It also denotes the ideal course in which to undertake the whole design process in a project, in which a real-life implementation and contribution to society are encouraged. In order to plan and realize the LeNSin Pilot Course as a Final Project course, there were thus three levels of objectives to be considered: the LeNSin Pilot Course particular objectives of designing for S.PSS applied to DE; the particular Final Project course objectives, and the objectives of the Undergraduate Design Programme.

Once the Pilot Course was defined to be integrated and planned as a one-year-long course, with a two-week "observation window", specific problems or thematic cases were established, as well as specific scheduling to accommodate the two campuses. calendar stages and topics related with the LeNSin Project. The thematic cases are briefly described in the following section.

6.2 Book Club and Desierto de Los Leones Park Projects

Libro Club (Book Club) is a government programme initially launched by the Cultural Ministry of Mexico City in 1998, whose main objective was to promote reading habits through the creation of open libraries (book clubs) managed by citizens throughout Mexico City. At the beginning of the programme, more than 1,019 book clubs were installed inside cultural and communitarian centres, hospitals, among others, having each one a basic bibliographic collection of around 500 books. The overall intention of book clubs was to offer a reading space, run by autonomous citizens, who could give unlimited access to their book collection through 'spoken agreement', based on trust. Due to an administrative change in the political agenda of Mexico City, much of the budget related to the Cultural Ministry was reduced and thereafter the Book Club programme became fragile and unstructured and more than 1200 clubs folded. The Book Club project on the UAM Cuajimalpa campus was thus an implementation opportunity for the Final Project and LeNSin Pilot Course, as a design intervention would provide an integrated strategy to strengthen and re-formulate book club structures from a systemic view, in a way that the resulting network would be authentically autonomous and resilient to all political changes through the redefinition of all its components as a socially relevant network.

The course structure consisted of three principal stages: contextual research and immersion through field study methods, development of design proposals, and proposal refinement and evaluation, in collaboration with the stakeholders, i.e. Book Club owners and users and representatives from the Cultural Ministry [16]. Synchro-nized with the general curricular course objectives, additional objectives were inter-woven in order to incorporate a methodological base that would allow students to:

- acquire an awareness related to the promotion of sustainable principles in emerging contexts;
- identify the social, economic and environmental spheres of a complex problem/system;
- understand the importance of the configuration process of stakeholders, interactions and scenarios throughout an interdisciplinary design process; and
- identify the theoretical and methodological basis of S.PSS and Distributed Economies.

Starting from the identification of the overall Book Club macro-system and its principal problems, stakeholders and sustainability challenges, students defined and structured their intervention through the articulation of sub-systems. This way, each team of students proposed specific S.PSS design strategies through autonomous, yet articulated, proposals. According to the specific identified problems, the overall proposed system included products and services that would allow Book Club owners and users to start, continue and self-manage an autonomous reading space.

On the UAM Azcapotzalco campus, the pilot course aimed to implement concepts and tools of social innovation to design, as well as a research process where the academic work of students was tied with those of research. Social innovation has been shown to be an important focus in sustainable design projects because it allows addressing the social variable from a novel perspective in the discipline: the user as a generator of their own solutions, where the designer reconsiders disciplinarity as a key part of the process. However, this poses new challenges for design education, since it implies a multidisciplinary practice that is not always affordable in the classroom. The objective of this project was to propose a product-service system that generates a significant change in social relations to improve the quality of life and employment of the community of vendors of the Desert of the Lions Park in the State of Mexico. The stakeholders involved in teaching and tutoring the students came from other disciplines, also from outside academia.

6.3 Summary

The LeNS pilot course, implemented for one year on both UAM campuses, provided a series of advantages at different levels. At one level, it was possible to bring students closer to the theories involved, related to both sustainability and the particular problems of each project, at a higher level of depth than what is possible in a two-week course. Students could be completely immersed in the problems, reaching important levels of empathy with the users and actors involved in the projects. Secondly, even though a general introduction was given to the different tools and principles of S.PSS Design and Distributed Economies during the two-week workshop [18], during the entire project they were introduced again during the appropriate phases of research, development and/or evaluation. These reviews of tools and principles were done in such a way that students had the time and opportunity to understand, analyse, test and execute them, not only at a conceptual level, but also in a real way once the prototypes and final proposals were developed.

Moreover, having the pilot course directly integrated into the curricula of the design programmes allowed the identification of the knowledge and skills necessary to cover in previous courses related to a deep reflection of sustainability, S.PSS design and economic paradigms (in the case of UAM Cuajimalpa), as well as the place and moment in which this knowledge could be distributed throughout the design programme, suitable teachers, contents, and so on. The implementation of a pilot course through the development of a real design project, in which the different stages of analysis, development and evaluation were transparent as evidence to the stakeholders, not only allowed a total commitment on the part of the students, but also reached a deep level of empathy and conviction of the methodological scope used.

7 Lessons Learned, Challenges and Opportunities

Besides the development of S.PSS design methods, the main success in the LeNSin project has been in the development of the network of educators that share an interest in developing teaching for sustainable design. During the project, several educators and students have collaborated in seminars and pilot courses, but also in thesis guiding, organizing seminars and events and faculty exchange. This development continues strongly from a shared history in design teaching and is well oriented to the shared challenges of today.

The project has allowed partners to study DE as a concept in various settings. Through the project, several design schools have connected to share experiences on the concept, and various actors and networks have been invited into collaboration. Within each pilot course, having lecturers from other universities acted as an alternative training process for future replications of the learning content. Since each professor has to deal with different local contexts, the result was a prolific field of discussions on how to implement S.PSS and DE methods and tools into design curricula. Another valuable outcome from the seminar and pilot interactions was the collection of several case studies from around the world on various DE interpretations. This work has continued through collaboration with selected local actors in pilot courses and in developing ideas for local DE solutions in various contexts of action in student case work.

During the project, it also became evident that S.PSS and DE as concepts are portrayed differently in different historical, geographical and political contexts. Discussing the emerging tensions can be of help in developing new content, forging collaboration and ensuring funding for future action. Understanding these dynamics is also of assistance in developing new interaction across the globe.

Sustainability is the grand challenge for the century and answers to its call are needed across professional fields. The design profession, as a potential matchmaker between different disciplines that are involved in the processes of planning and development, also calls for new methods and tools to create new interpretations of more sustainable solutions. S.PSS and DE as approaches to design can also provide new perspectives on social sustainability, extending the considerations in conventional eco-design.

In solving the challenges of the twenty-first century, future designers need to become change agents and help to expand sustainability considerations further. In this process, projects such as LeNSin—and the networks that can be developed through them—are crucial mechanisms to take work further, to legitimize action across various settings and actors.

References

1. Adams R, Martin S, Boom K (2018) University culture and sustainability: designing and implementing an enabling framework. J Clean Prod 171:434–445
2. Banerjee S, Upadhyay P, Punekar RM (2019a) Teaching design for sustainability for socioeconomic ecosystems—three case studies. In: Research into design for a connected world, pp 935–946
3. Banerjee S, Upadhyay P, Punekar RM (2019b) Contextualising sustainable product-service design methods for distributed economies of India. In: Ambrosio M, Vezzoli C (eds) Designing sustainability for All—Proceedings of the 3rd LeNS world distributed conference. Edizioni Poli.design, Milano, IT, pp 270–275
4. Boradkar P (2010) Design as problem solving. In: Frodeman R (ed) The Oxford handbook of interdisciplinarity. Oxford University Press, Oxford, UK, pp 273–287
5. Brown VA, Harris JA, Russell JY (eds) (2010) Tackling wicked problems through the transdisciplinary imagination. Earthscan, US-DC, London, UK; Washington
6. Ceschin F, Gaziulusoy I (2020) Design for sustainability—A multi-level framework from products to socio-technical systems. Routledge, New York
7. Cross N (2011) Design thinking: understanding how designers think and work. Bloomsbury, New York, NY
8. Frodeman R (ed) (2010) The Oxford handbook of interdisciplinarity. Oxford University Press, Oxford, UK
9. Gaziulusoy Aİ, Erdogan Öztekin E (2018) Design as a Catalyst for Sustainability Transitions. DRS 2018. https://doi.org/10.21606/dma.2017.292
10. Jégou F, Manzini E (2008) Collaborative services: social innovation and design for sustainability. Edizioni Poli.design, Milano, IT
11. Kohtala C, Hyysalo S, Whalen J (2020) A taxonomy of users' active design engagement in the 21st century. Des Stud 67:27–54
12. LeNSin Project (2019) Student design competition catalogue 2019. http://lensconference3.org/images/program/LeNS_competition_catalogue_2019.pdf
13. Manzini E (2019) Sustainability and democracy—Widespread collaborative design intelligence [presentation]. Designing Sustainability for All: The LeNS Distributed Conference 2019. Milan, Italy, 3–5 April
14. Marttila T (2018) Platforms of co-creation: learning interprofessional design in creative sustainability. Doctoral dissertation. Aalto University, Espoo, FI
15. Phukan R (2012) Muga silk industry of Assam in historical perspectives. Global J Hum-Soc Sci Res 12(9-D)
16. Sagahon L, García B (2019) Introducing systemic solutions for sustainability at design courses in UAM Cuajimalpa. Study case: Book Club in Mexico City. In: Ambrosio M, Vezzoli C (eds) Designing sustainability for all: proceedings of the 3rd LeNS world distributed conference. Edizioni POLI.design, Milan, Italy, pp 16–20
17. Saikia JN (2011) A study of the Muga silk reelers in the world's biggest muga weaving cluster-Sualkuchi. Asian J Res Bus Econ Manag 1(3):257–266
18. dos Santos A, Vezzoli C, Kohtala C, Srinivasan A, Diehl JC, Fusakul SM, Xin L, Sateesh D (2018) Sistema Produto+Serviço Sustentável: Fundamentos. Insight, Curitiba, Brazil

Printed in the United States
by Baker & Taylor Publisher Services